Encounter
Human Geography

INTERACTIVE EXPLORATIONS OF EARTH
Using Google Earth™

JESS C. PORTER

PEARSON

Boston   Columbus   Indianapolis   New York   San Francisco   Upper Saddle River
Amsterdam   Cape Town   Dubai   London   Madrid   Milan   Munich   Paris   Montréal   Toronto
Delhi   Mexico City   São Paulo   Sydney   Hong Kong   Seoul   Singapore   Taipei   Tokyo

Geography Editor: Christian Botting
Marketing Manager: Maureen McLaughlin
Assistant Editor: Kristen Sanchez
Marketing Assistant: Nicola Houston
Editorial Assistant: Bethany Sexton
Managing Editor, Geosciences and Chemistry:
 Gina M. Cheselka
Project Manager, Production: Edward Thomas
Media Producer: Ziki Dekel
Supplement Cover Designer:
 Seventeenth Street Studios, Richard Whitaker
Operations Specialist: Maura Zaldivar
Senior Operations Manager: Nick Sklitsis

Front cover image credit:
 Earth globe courtesy of NASA.
Back cover image credits:
 (Left) ©2011 MapLink/TeleAtlas, ©2011 Europa
 Technologies, US Dept of State Geographer.
 (Right) ©2011 Lead Dog Consulting, ©2011
 Europa Technologies, © 2011 Digital Globe, ©2011
 GeoEye.
Inside front cover image credits:
 (Top) ©2011 TeleAtlas, ©2011 Cyber City, ©2010
 VirtualCity, ©2011 Aerodata International Surveys.
 (Bottom) USDA Farm Service Agency, ©2011
 Digital Globe, ©2011 Google™ , ©2011 GeoEye.

Printed in the United States of America

10 9 8 7 6 5 4 3 2 1

ISBN 10:   0-321-68220-3
ISBN 13: 978-0-321-68220-8

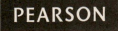

Contents

Preface

Welcome to *Encounter Human Geography*! This workbook will immerse you in interactive explorations of the world's immense human geographic diversity. Elements of human geography and geospatial techniques come together to give you a better understanding of our worlds cultural variations and areas of common ground. This is accomplished by applying the power of the Google Earth™ program to zoom in and around features and landscapes ranging from street-corner scenes to regional-scale economic patterns. Within Google Earth™, we will utilize associated tools and layers such as photographs, satellite imagery, and historic maps. We will springboard into related websites that help us understand the patterns and processes taking place on Earth and among its diverse peoples.

As you work through the exercises contained in this book, you will feel your "big picture" understanding of the world come into focus. Not only will your knowledge of the themes of geography grow, but also your ability to apply these themes to your interpretation and understanding of life on Earth. While these exercises are designed to educate thematically and topically, they are also designed to be fun. We encourage you to take these exercises beyond the parameters laid out in the multiple choice and short answer question segments of the workbook. If something piques your curiosity, dig-in deeper. Look for the answers, but also look for more questions to ask. It is our hope that you will apply your enhanced spatial understanding of the world and its cultural attributes beyond your studies associated with this workbook. You will find that improving your spatial thinking skills is something that can benefit every aspect of your life.

Google Earth™ can be downloaded for free from http://earth.google.com. The Google Earth™ web site has a wide variety of instructional materials from tutorials to demonstration videos. You can also download interesting collections of spatial data in KMZ or KML files. Explore the software and take advantage of Google's ancillary material to familiarize yourself with your new aid to geographic learning.

This workbook is organized with two introductory chapters. The first provides you with a Google Earth™ primer as you learn to navigate and use the software's system of layers to display spatial information. The second chapter addresses some key concepts in geography that are essential for you to get the most out of this workbook. Location, scale, and place are discussed and the basics of interpreting remotely sensed images are introduced. This will help you better understand and interpret what you are seeing in the Google Earth™ environment. The workbook then examines 17 elements of human geography beginning with population.

Each of these chapters examines four subtopics where you will have the opportunity to answer multiple-choice and short answer questions. These questions have been designed with an emphasis on high-level assessment skills that will encourage you to interpret and appraise the imagery and information that you uncover. You will utilize Pearson's companion website, www.mygeoscienceplace.com, to download .kmz files, worksheets, and electronic versions of the assessments.

Finally, I want to thank the editors and staff of Pearson Geography for all of their hard work, patience, and flexibility in the development and production of *Encounter Human Geography*. Specifically, it was great to team with two talented, intelligent, and creative collaborators, Christian Botting and Kristen Sanchez.

Jess C. Porter
jcporter@ualr.edu

Correlation Grid

Encounter Human Geography Porter	The Cultural Landscape Rubenstein	Human Geography Knox and Marston	Contemporary Human Geography Rubenstein
1	1	1	1
2	1	1,6	1
3	2	3	2
4	2	3	2
5	3	3	3
6	4	5	4
7	5	5	5
8	6	5	6
9	7	5	7
10	2,9	5,6	9
11	8	9	8
12	8	9	6,8
13	9	1,2,7	9
14	10	8	10
15	11	7	11
16	12	7	12
17	13	10,11	12,13
18	14	4,7	14
19	14	4	14

Name:_____

Date: _____

Chapter 1:

Introduction to Google Earth™

This chapter will introduce you to Google Earth™. Google Earth™ is a computer application that provides opportunities to see the Earth's varied geography from a range of aerial and street-level perspectives. You can zoom in and out, rotate, and tilt your perspective of the Earth. You will learn how to navigate to different locations and use some of the tools that this powerful application provides. Note: Google Earth™ is an application that is updated frequently. As a result, the functions and appearance of the application may vary slightly from what is described below.

To begin, verify that your computer has the latest version of Google Earth™ installed. If you do not, then go to http://earth.google.com and download the most recent version of the free Google Earth™ software. Your computer will need a high-speed internet connection in order to utilize the application effectively.

The very best way to learn Google Earth™ is to dive in and explore your world. The software is intuitive to use and not complicated. However, you will also find it useful to learn some of the basics of navigation as demonstrated in the Google Earth™ User Guide. Go to http://earth.google.com/support and then click the link for "User Guide." Do not forget about this useful resource. If you are ever confused or need help with using the software, this will be your best bet for quick and easy assistance.

In the User Guide, under the "Getting started" heading, there is a link titled "Getting around." Open this link and study the "Getting to Know Google Earth™" diagram and associated information. Be sure to click the links associated with the tools to explore the capabilities of Google Earth™. Without studying the aforementioned pages of the User Guide, you will be unable to answer the following questions and will likely encounter significant difficulties in your attempts to complete the explorations of Encounter Human Geography.

Download EncounterHG_ch01_Intro.kmz from **www.mygeoscienceplace.com** *and open in Google Earth™.*

Exploration 1.1: Getting Started

1. If you needed to find a specific place on Earth, you would use the

   A. Places panel.
   B. Layers panel.
   C. Overview map.
   D. Finder tool.
   E. Search panel.

2. The on-screen navigation controls for Google Earth™ are located on what part of the Google Earth™ window?

   A. bottom right
   B. top right
   C. center
   D. bottom left
   E. top left

3. Which of the following is not a capability of Google Earth™?

   A. You can display sunlight across the landscape.
   B. You can add polygons, lines, and points (placemarks) to the view.
   C. You can view live imagery.
   D. You can display historical imagery.
   E. You can email views from Google Earth™.

4. Where would you find the information regarding a view's coordinates, elevation, and imagery date?

   A. In the Places panel
   B. In the Search panel
   C. In the Layers panel
   D. At the bottom of the 3D Viewer
   E. In the Overview map

In the User Guide, click the "Navigating in Google Earth™" link, located under the "Getting around" heading. Play the short video to learn how to navigate the globe in Google Earth™. Scroll down the page and review other tips for navigating in Google Earth™. Be sure your computer's sound is turned on.

5. If you wanted to look around from one vantage point, as if you were turning your head, you would use the

   A. on-screen look joystick.
   B. arrows on your keyboard.
   C. on-screen move joystick.
   D. wheel on your mouse.
   E. on-screen zoom slide.

A helpful feature of Google Earth™ is the Overview Map. This shows where you are on the globe when you are zoomed in closely to the Earth. Click the link to "Using the Overview Map" in the "Getting around" section of the User Guide.

6. Identify the statement regarding the Overview Map that is supported by the information contained in the User Guide.

   A. The zoom ratio of the Overview Map cannot be adjusted.
   B. The Overview Map window responds to adjustments made in the 3D Viewer.
   C. The Overview Map is always shown in Google Earth™
   D. The Overview Map is intended to show areas at a larger scale than the 3D Viewer.
   E. The size of the Overview Map is larger than the 3D Viewer.

Click the "Five Cool, Easy Things You Can Do in Google Earth™" link, located under the "Getting around" heading. Number one on the list explains how to enter the location of your school in the Search panel and then view the image of your school. Do this now.

Exploration 1.1: SHORT ESSAY

1. Seeing your school from an aerial perspective, describe some insight that you have gained. Did you find out there were buildings you did not know existed? Maybe there is a parking lot that is closer to your dorm than the one you currently use? Feel free to be creative in your response.

_____
_____
_____
_____
_____
_____

One can navigate in Google Earth™ using either on-screen controls, the mouse, the keyboard, or any combination of these approaches. To see what keyboard commands perform what tasks, click the "Navigating with the keyboard" link in the "Getting around" section of the User Guide. Note that the keystrokes differ for a Mac and a PC.

2. Navigate using both the on-screen controls such as the zoom slide and move joystick; then navigate using the keyboard and/or mouse. Try integrating the two approaches. Which method(s) do you prefer and why?

_____
_____
_____
_____
_____
_____

Exploration 1.2: Layers

Google Earth's™ functionality is magnified by the ability to add layers of data and information to the 3D Viewer. Beyond the visual representation of the Earth's surface we can add any number of features such as the road networks, place names, boundaries, photographs, business locations, current weather conditions, and three-dimensional buildings to name just a few. A problem that faces many users of Google Earth™, however, is that their viewers become cluttered with too much information. In general, it is best practice to simply display only the layers you need at any given time. *Borders and Labels* is the layer that you will find most useful without cluttering the view.

Open Google Earth™ and begin by going to the Layer pane and turning off all of the layers. This can be done with a single click by un-checking the box next to *Primary Database*.

Open the *1. Introduction to Google Earth™* folder that is loaded in the *Temporary Places* folder in the Places pane. You can right-click on the *Introduction to Google Earth™* folder and you will see an option to "Save to My Places." Select this and the folder will be moved to your *My Places* folder. This way you will not have to reload it next time you start Google Earth™.

By clicking the plus sign next to the *1. Introduction to Google Earth™* folder, you will expand the folder and see what's inside. You will see a series of five folders. Expand the *1.2 Layers* folder and then double-click the *Albuquerque, NM* placemark. Now check the boxes beside the ten or so layers and/or folders in the *Primary Database* in the Layers pane beneath the Places pane. Wait a few moments and you will get a taste of how cluttered Google Earth™ can become. It will take several moments for everything to load. Now turn off all of the layers/folders except the folder called *Borders and Labels*. Depending on the type of question you seek to answer, you may want to activate one or several folders or layers at a time.

For example, let's start with the *Photos* layer. Turn this layer of information on by clicking the box next to *Photos* in the *Primary Database*. You will see a number of small blue and red squares appear on the screen. Begin clicking the blue squares and you will see that these are photographs of features that are correlated with their location on the ground. Click on approximately ten of these boxes contained in the current view. Now click one of the red squares to see 360-degree views that are available. When these images are activated, you can click the photograph to fly into the image. Click "Exit Photo" in the upper-right corner to return to the 3D Viewer. Return to the original perspective by double-clicking the *Albuquerque, NM* placemark.

Turn off all of the layers except the *Roads* layer. You will see that the major highways of the community are illuminated and labeled. This *Roads* layer can be very helpful when you are attempting to analyze patterns on the landscape, as transportation is closely related with many types of development. Now use the zoom-slider or the scroll wheel on your mouse to slowly zoom in. As you zoom in, you will notice that more and more details of the road network will emerge until the smallest residential streets appear and are labeled. Zoom out and these features and labels disappear.

Exploration 1.2 MULTIPLE CHOICE

1. The name of the street that is immediately north of the *Albuquerque, NM* placemark is

   A. Indian School Rd.
   B. Interstate 40.
   C. State Highway 47.
   D. Richmond Dr.
   E. Euclid Ave. NE.

Turn off the *Roads* layer. Double-click the *Downtown* placemark in the Places pane to zoom to a view of downtown Albuquerque. Perhaps you can get a sense that this is the central business district by the relatively larger buildings. Maybe you noticed the shadows in the image suggesting some of these buildings had considerable vertical development. It would be helpful if we had more information that could be added to the map to help us understand the physical realities of downtown. Turn on the *3D Buildings* folder and wait a moment for the data to load. You can click on individual 3D models that have been built to gain additional information. Now you have a much better feel for what the downtown area really looks like.

This added information is what geographic information systems (GIS) are all about. We can take layers such as a road network and do analysis in relation to the built environment. Throughout this text you will be capitalizing on the ability of Google Earth™ to display and overlay different types of data and information on the surface of the digital globe in order to increase your understanding of the world and human geography.

2. What is the name of the tallest building in downtown Albuquerque?

   A. New Mexico Bank and Trust
   B. Albuquerque Convention Center
   C. Commerce Bank of Albuquerque
   D. Albuquerque Plaza Office Tower/Bank of Albuquerque
   E. Albuquerque Petroleum Building

Turn off the *3D Buildings* folder. Street-view imagery is increasingly available in urban locations around the world. This tool can really do a great job of providing you with a sense of place of a particular location as you can simulate driving down a road by clicking on successive camera icons. From the perspective of the *Downtown* placemark, locate Central Avenue near the yellow push-pin for downtown. Continue to zoom in toward the street near the placemark until the screen switches to Street-view. Work your way down Central Avenue to the 300 block address range. You can advance down the street by clicking on the yellow line or using your mouse's scroll wheel. The address is shown in the upper right corner of the screen. There is a bar and grill with a unique sign in the aforementioned address range.

3. The bar and grill that is located in the 300 block of Central Avenue has a unique sign that identifies it as the

    A. Library Bar and Grill.
    B. Eagle and Child Bar and Grill.
    C. Ritz Bar and Grill.
    D. Long Bar and Grill.
    E. Celtic Bar and Grill.

Click "Exit Street View" in the upper-right corner when you have completed your street-viewing. Turn on and double-click the *US Drought 04/09/11* folder to see an illustration of the application of areal data to Google Earth™. This shows drought conditions on April 09, 2011. Here it would be helpful to know what states are affected, so let's turn on the *Borders and Labels* folder. Sometimes it's advantageous to make a data layer somewhat transparent. At the bottom of the Places pane you will see a slider that will allow you to adjust the transparency of an overlay. Practice manipulating the transparency before returning the slider to the maximum coverage (slid to the right).

4. The USDA drought imagery illustrates that drought conditions were consistently most severe in early April of 2011 in

    A. Minnesota.
    B. the Pacific Northwest.
    C. an area from Maine to New York.
    D. south Texas.
    E. the Dakotas.

Turn off the *US Drought 04/09/11* folder. In the *Primary Database*, expand the *Weather* folder. Turn on the *Clouds* layer and explore the coverage of this data layer and then turn on the *Radar* layer. You will see that these layers do not have the same coverage. While cloud imagery is available globally, the radar data is only available for select locations like North America and Western Europe. This data is one of the rare occasions in Google Earth™ where you are viewing near-live data. When you have completed your exploration of clouds and precipitation, turn off and collapse the *Weather* folder.

There are dozens of additional datasets, multimedia files, and links to outside resources in Google Earth™. For a taste, expand the *Gallery* folder in the *Primary Database*. Turn on the *NASA* folder and then navigate to southern Ethiopia. Click on the NASA icon and explore the NASA content that is available for this location.

5. What statement best describes the NASA data available associated with the NASA icon positioned over southern Ethiopia?

    A. Volcanic activity that is ongoing in the region.
    B. Effects of Saharan dust storms on Ethiopia's cities.
    C. Imagery of vegetation deviation from normal.
    D. Pollution from industrial fires.
    E. Copper mining impacts on the landscape.

Exploration 1.2 SHORT ESSAY

1. Based solely on what you can ascertain from your sampling of Panoramio photographs, write a short paragraph that describes the community of Albuquerque, New Mexico.

_____
_____
_____
_____
_____
_____

2. Describe the clouds and precipitation patterns over North America using the Radar data from Google Earth™. Is there a strong correlation between cloud cover and precipitation? Can you identify a storm system(s) (area(s) of low pressure)? In other words, does it look like it could be raining in or near your area right now? How can you tell? Document the date and time you viewed these features in your response.

_____
_____
_____
_____
_____
_____

Exploration 1.3: Human-Environment Geography Elements

Let's continue to build our familiarity with Google Earth™ and its tools by exploring a few locations that introduce some of the topics of human geography. As you complete the explorations, you will find that these topical areas overlap and intertwine with great regularity. Although our focus is on the human elements of geography, the physical landscape is usually a significant component of our studies. In fact, Google Earth™ is a terrific tool to visualize Earth's complex physical landscapes. The three-dimensional capacity of the program allows the user to get a grasp on the variations in relief that a paper-map is not able to provide. In the *Primary Database*, verify that all layers are turned off.

Open the "Google Earth Options" dialog box by selecting "Tools," and then "Options" from the menu bar. On the "3D View" tab, verify that elevation is displayed using "Meters, Kilometers" and click "OK." Open the *1.3 Human Geography Elements* folder and double-click the *Grand Canyon* placemark. You will zoom to a view of this unique natural feature. As you move the cursor across the screen you will see the values for elevation that are displayed at the bottom of the screen change.

Exploration 1.3: MULTIPLE CHOICE

1. What is the approximate elevation value of the Colorado River in the vicinity of the *Grand Canyon* placemark?

   A. 725 meters above sea level
   B. 1145 meters above sea level
   C. 1550 meters above sea level
   D. 2375 meters above sea level
   E. 6335 meters above sea level

Turn on the *Deepwater Horizon* placemark and then double-click it to fly to the location. This is the site of the 2010 oil spill that occurred after a deep-sea drilling rig experienced a total failure. Let's use the measure tool (the button containing a ruler icon at the top of the screen) to determine the distance between the site of the accident and the nearest shoreline. Click the Ruler tool and then position the cursor over the *Deepwater Horizon* placemark. Click once and the move the cursor over the nearest land (the end of the Mississippi River delta) and click again. Be sure you have changed the units to kilometers using the drop-down menu associated with Length in the Ruler tool.

2. What is the <u>approximate</u> straight-line distance between the *Deepwater Horizon* site and the nearest land?

   A. 20 kilometers
   B. 50 kilometers
   C. 80 kilometers
   D. 120 kilometers
   E. 175 kilometers

An important part of monitoring the spill in the days, weeks, and months after the event was the application of satellite imagery. Turn on the *2010-05-17 MODIS* folder and zoom to an appropriate level to view the extent of the spill.

3. At the time the *2010-05-17 MODIS* image was captured, the oil spill extended farthest from the *Deepwater Horizon* site in the

   A. western direction.
   B. northern direction.
   C. eastern and northern directions.
   D. northern and western directions.
   E. southern and eastern directions.

Now turn on and double-click the *New Orleans, LA and Suburbs Flood Zones* layer. This map has been used as a guide for residents of New Orleans who decided to rebuild following Hurricane Katrina. Notice the slider at the bottom of the Places pane. You can use this to make the active layer (the one that is highlighted in the Places pane) more or less transparent. Experiment with the slider now.

4. The area of New Orleans that is generally lowest in elevation is

    A. the Algiers neighborhood, south of the Mississippi River.
    B. the area south and east of the Lakefront Airport.
    C. downtown.
    D. the area due east of Louis Armstrong International Airport.
    E. the Elmwood neighborhood, north of the Mississippi River.

Exploration 1.3 SHORT ESSAY

1. Based on what you have seen regarding the Deepwater Horizon oil spill and the rebuilding of New Orleans, identify and describe two ways Google Earth™ can be applied as a beneficial tool in times of crisis.

    _____
    _____
    _____
    _____
    _____
    _____
    _____

One extremely useful capability of Google Earth™ is the ability to view historical imagery for many locations on Earth. This allows one to document change over time. Double-click the *Japan Tsunami* placemark. You will see one small area that was devastated by the tsunami that occurred following an earthquake on March 11, 2011. You will notice that a time slider has appeared in the upper-left corner of the 3D Viewer. This can be turned on and off by clicking the historical imagery tool. It has a picture of a clock with an arrow pointing in the counter-clockwise position. With the historical imagery turned on, click the rewind button on the far left of the slider to step back one image set at a time. You can cycle back until you see pre-tsunami imagery. When you have completed the following question, be sure to turn off the historical imagery.

2. What is the most recent date that pre-tsunami imagery is available for this location? Describe some of the changes you can see in the setting from the time of the tsunami up to the most recent imagery.

    _____
    _____
    _____
    _____
    _____
    _____
    _____

Exploration 1.4: Human Geography Elements

Let's take a look at a sample of population data displayed in Google Earth™. Double-click and turn on the *Life Expectancy Birth* folder. The height and color of the country polygons correlate to the life expectancy values. The countries that are more extruded (taller/higher) have values that are higher. Colors are ramped to illustrate locations at the top and the bottom of the scale. You can click on the flag of a country to see its value, as well as the highest and lowest value in the dataset. Turn off the *Life Expectancy Birth* folder when you complete the following question.

1. Which of the following countries has the lowest life expectancy at birth?

    A. Afghanistan
    B. Japan
    C. Laos
    D. United States
    E. Zambia

Our sampling of human phenomenon continues by double-clicking and turning on the *USA Religious Adherents* layer. Counties are enumerated according to the percentage of persons per county that adhere to a religious body. Assess the map for patterns. Turn off the *USA Religious Adherents* layer when you complete the following question.

2. Which of the following statements is <u>not</u> supported by the *USA Religious Adherents* layer?

    A. Utah has one of the highest state rates of religious adherence.
    B. Oregon displays one of the highest levels of religious adherence.
    C. The east and west coasts generally have lower levels of adherence than the interior of the US.
    D. Religious adherence in the Great Plains is generally higher than in the Rocky Mountains.
    E. The map suggests that Illinois is more religious than Indiana.

Now view the popular culture location associated with the *Cultural Icon* placemark. Double-click this placemark. Study the site and the surrounding area and use clues provided by other layers you can turn on in the *Primary Database*. When you complete the following question, turn off any layers that you have turned on in the *Primary Database*.

3. The site associated with the cultural icon placemark is

    A. Las Vegas.
    B. the Mall of America.
    C. Disneyland.
    D. the state capital of Arizona.
    E. the Neverland Valley Ranch.

Within the aforementioned cultural icon site, there is at least one building that could be identified by many people around the world. The built environment can provide us with buildings that reflect both folk and popular culture and the values of a local society in general. The next folder, *Unusual Buildings*, contains a fun collection of interesting structures from around the world. Click on the placemarks in the map or in the Places pane to see pictures of these structures.

4. In what state is found a unique building that suggests fishing is a significant part of the local culture? Remember you can turn on the Borders and Labels layer to see the states labeled.

    A. California
    B. Georgia
    C. Maryland
    D. New York
    E. Wisconsin

Exploration 1.4: SHORT ESSAY

1. How does your perception of the site associated with the *Cultural Icon* placemark change after viewing it from this perspective? For example, is it bigger or smaller than you expected? Were you surprised by the surrounding landscape?

    _____
    _____
    _____
    _____
    _____
    _____
    _____

Turn on and double-click the *AirTraffic 3D* folder. This is a neat application that tracks airline flights in central Europe in real time. Observe the screen for a few minutes and note any patterns you see with trajectory and altitude of planes.

2. What kind of information can you obtain from the *AirTraffic3D*? What real-world benefits could this or similar tools that have been created for Google Earth™ provide?

    _____
    _____
    _____
    _____
    _____
    _____
    _____

Name: _____

Date: _____

Chapter 2:
Geography Concepts

As you work through the exercises contained in this workbook, you will become well-versed in some of the basic concepts of the discipline of geography. You will hone your understandings of location, scale, and place by analyzing remotely sensed data and information. The ability to effectively interpret this data and information is the key skill these exercises seek to develop.

Download EncounterHG_ch02_Geography_Conceps.kmz from www.mygeoscienceplace.com and open in GoogleEarth™

Exploration 2.1: Location

Location is a concept that can be thought of in absolute or relative terms. For example, you are sitting in front of your computer is an example of relative location. On the other hand, absolute location refers to any system that provides us with a set of coordinates that represent a specific point on Earth's surface. The most commonly used method for determining location is the grid system of latitude and longitude. Latitude measures distance north and south of the Earth's equator, while longitude provides a measure of distance east and west of the Prime Meridian.

Open the *2. Geography Concepts* folder and then open the *Location* folder and turn on and double-click the *Point of Origin* placemark. This is the origin for our globe's predominant location system. In other words, this is where the Earth's east/west baseline (the Prime Meridian/International Date Line) and the Earth's north/south baseline (the equator) intersect. Let's make it a little easier to see. Click "View" and then click "Grid." The latitude and longitude grid is illuminated. You can see that the point of origin placemark is located at 0° latitude and 0° longitude.

Now let's make sure that we are on the same page in terms of the way latitude and longitude is displayed. Click "Tools," and then "Options." The Google Earth™ Options dialog box will open. In the "Show Lat/Long" box, you will see four options for displaying your absolute location. Select "Decimal Degrees" and then click "OK." At the center bottom of your screen you will see the coordinates for the location of your cursor (the white hand). Move it around the globe. Examine what happens to the latitude and longitude numbers as you move to the southern and northern hemispheres and the eastern and western hemispheres. A positive latitude number represents locations north of the equator while a positive longitude represents locations east of the Prime Meridian. Move your cursor across the grid again to verify these statements.
Now manipulate the globe by rotating it and turning it to explore latitude and longitude at the poles and in the Pacific Ocean. Determine the maximum values for latitude and longitude and where those are located.

Exploration 2.1: MULTIPLE CHOICE

1. Which of the following coordinate pairs is associated with Lisbon, Portugal?

   A. 38° North latitude, 7° West longitude
   B. 38° South latitude, 7° West longitude
   C. 38° North latitude, 7° East longitude
   D. 38° South latitude, 7° East longitude
   E. 7° North latitude, 38° West longitude

2. Based on your assessment of latitude and longitude, which of the following statements is true?

   A. The maximum value of longitude is 360°.
   B. The North and South Poles are located at 90° north and south latitude.
   C. All locations south of the equator have positive values of latitude.
   D. All locations located west of the Prime Meridian and east of the International Date Line (Antemeridian) will have east longitude.
   E. The equator represents the 0° line of longitude.

3. Based on your assessment of latitude and longitude, which of the following statements is true?

   A. Lines of longitude are parallel with one another, thus insuring that a degree of longitude is equidistant to all other degrees of longitude anywhere on the globe.
   B. All of Europe is classified as west latitude.
   C. New Zealand's longitudinal values are higher (more eastern) than Australia's.
   D. The Prime Meridian represents the 0° line of latitude.
   E. Lines of latitude converge with one another at the poles.

4. What physical feature is located at -15.72° latitude, 29.35° longitude?

   A. mountain
   B. island in the ocean
   C. river channel
   D. peninsula
   E. volcano

Exploration 2.1: SHORT ESSAY

1. Use Google Earth™ to determine the latitude and longitude of your school and your home. List the coordinates for both and explain why the numbers are lower or higher in comparison with one another.

_____

_____

_____

_____

_____

_____

_____

2. Use Google Earth™ to locate three places you want to visit in your lifetime. List each place, its country, its latitude and longitude, and whether it is in the northern or southern hemisphere and the eastern or western hemisphere.

_____

_____

_____

_____

_____

_____

Exploration 2.2: Scale and Place

Another important concept to grasp is that of scale. When we use the term _scale_, we are referring to the relationship between units on a map or, in this case, a digital globe, compared to units in the real world. An example of a verbal scale that you may have seen before would be "1 inch equals 1 mile." A scale can also be portrayed as a fraction, such as 1:1000. This means that one unit on the map is equal to 1000 units in the "real world." Large-scale maps show a large amount of detail but not much area, whereas a small-scale map shows a more extensive area with less detail. Therefore, a 1:1000 scale map is a larger scale map than a 1:5000 scale map. As you will experience in your work with Google Earth™, some types of analysis are better suited to viewing the Earth at larger scales (zoomed in more), while others dictate a small-scale approach (zoomed out more).

Place is a final essential concept. Place refers to what makes one location unique from another. Place analysis is a subjective endeavor as the geographer tries to illuminate idiographic characteristics of a location. Oftentimes, these characteristics are based on social and cultural attributes such as economy, language, ethnicity, and religion.

Turn off the latitude longitude grid to unclutter the screen, but remember you can always turn it back on if you find it helpful or necessary to solve a problem. Double-click the _Mt. Kilimanjaro_ placemark. You'll see that Mt. Kilimanjaro is located in the vicinity of -3° latitude, 37° longitude. As a famous mountain, it would be helpful to get a grasp on the elevation of the mountain. When you move your cursor across the landscape you will see that the values for elevation displayed at the bottom of the screen will vary. Notice how the elevation decreases as you move away from the summit of Kilimanjaro. Zoom in and identify the highest point on Mt. Kilimanjaro. Remember that you will need to have the "Meters, Kilometers" option set for displaying elevation.

Exploration 2.2: MULTIPLE CHOICE

1. The summit of Mt. Kilimanjaro exceeds

   A. 15,000 meters.
   B. 7,300 meters.
   C. 5,800 meters.
   D. 2,900 meters.
   E. 1,500 meters.

Now double-click the *Mt. Kilimanjaro 2* placemark. Notice that this perspective of the same location gives you a better visual appreciation of the relief in the area of Mt. Kilimanjaro. Sometimes it might be helpful to exaggerate the variations in elevation in order to see patterns on the landscape. Click "Tools," and then "Options." The Google Earth™ Options dialog box will open again. In the "Terrain Quality" box change the "Elevation Exaggeration" to "3" and then click "OK." You will need to back out to see all of Kilimanjaro. Beyond the mountain being more dramatically represented, notice how the stream channels on the flanks of Kilimanjaro are now more visible. Rotate the scene 360° by dragging the "N" in the compass. You will see other prominent mountains. Find the highest peak within approximately 100 kilometers. Remember that the Ruler tool can help you determine a 100 mile radius around Kilimanjaro.

2. The highest peak in the vicinity (<100 kilometers) of Mt. Kilimanjaro is located to the

   A. north.
   B. south.
   C. east.
   D. west.
   E. northeast.

Return to the Google Earth™ Options dialog box and reset the "Elevation Exaggeration" to "1." Double-click the *Denver 1* placemark. You see the greater Denver metropolitan area. It's difficult to pick out many features of the metropolitan area at this scale. You should be able to identify the mountains to the west of town, some of the stream networks, agricultural regions, and the urban area. This is the smallest scale view of Denver that we will utilize.

Now double-click the *Denver 2* placemark. This view is a larger-scale view than the one associated with *Denver 1*. By zooming in we can see much more detail, but we've lost some of the big picture. For example, the high mountains are no longer visible. But now we can see highways and the major roads and features such as reservoirs and green space. Move to the *Denver 3* placemark. Much has been gained by zooming in. This is a larger-scale view than *Denver 2*, but will be smaller scale than *Denver 4*. Now we see features such as sports stadiums, highway interchanges, and the downtown area. However, we have no idea how big the metropolitan area is and we have no way of knowing that there are mountains immediately to the west.

3. What can be verified by the *Denver 3* view?

    A. The major transportation corridor visible generally runs east/west.
    B. The downtown area street grid is set on a 45° angle.
    C. There are no large football and baseball stadiums in the view.
    D. No major streams are seen in the vicinity of the primary transportation corridor.
    E. The major transportation corridor has many bridges in the view.

Double-click *Denver 4* and you will see a level of detail that enables you to gain an in-depth understanding of the features on the ground. Here we see an amusement park and its parking lot along with a river. Large-scale images like this facilitate analysis and interpretation. For example, we could use this image to determine the capacity of the parking lot, the spatial distribution of crosswalks, or the amount of the park that was shaded by trees.

This rich detail begins to give us a sense of this place. Clearly, this is a society with a relatively high level of affluence as evidenced by the abundance of recreational facilities. Google Earth™ has a number of layers available that can help us develop a sense of place. One of the best is the 360° panoramic images. Fly into the *Denver Street-view*. Manipulate the view to look up, down, and all-around.

4. Based solely on the *Denver Street-view*, what conclusion is supported?

    A. Mass transit is available.
    B. Vehicular transportation is not evident.
    C. There is likely a Hindu population in the immediate area.
    D. The climate here is extremely wet.
    E. Property in this vicinity is not very valuable as evidenced by the vertical development.

For a comparison, fly into the *Shibam Street-view*. Clearly this urban snapshot of life in Yemen is a vastly different scene than what you explored in the *Denver Street-view*.

5. Based solely on the *Denver* and *Shibam Street-views*, what conclusion is not supported?

    A. The climate in Shibam is likely arid.
    B. The Shibam image is an area visited by tourists.
    C. Infrastructure such as roads and the electrical grid is less developed in Shibam.
    D. Some women in Shibam wear very modest attire.
    E. There is an emphasis on modern architecture in Shibam at the expense of folk/traditional architecture.

Exploration 2.2: SHORT ESSAY

1. Return to the *Denver 4* placemark and study the image. Provide three additional questions that could be explored with this image (e.g., what's the capacity of the parking lot?).

   _____
   _____
   _____
   _____
   _____
   _____
   _____

2. What do you see in the *Shibam Street-view* that is most different to your daily life and what is most similar?

   _____
   _____
   _____
   _____
   _____
   _____
   _____

Exploration 2.3: Remotely Sensed Data

The imagery that you view in Google Earth™ has all been captured from remote sensors such as special cameras attached to satellites or airplanes. For the purpose of this text, it's not critical that you understand the intricacies of the sensors involved, but rather that you are aware of the nuances of remotely sensed imagery that is utilized by Google Earth™. This exploration will help you understand some of the issues you are likely to encounter when working with Google Earth™.

Open the *2.3 Remotely Sensed Data* folder and double-click the *Resolution 1* placemark. "Resolution" refers to the spatial resolution or the measurement of the minimum distance between two objects that will allow them to be differentiated from one another in an image. In effect, resolution can be thought of as the degree of clarity of the image. A higher resolution image would have a lower number associated with it. For example, a 30-meter spatial resolution image would be a higher resolution image than a 100-meter spatial resolution image. The imagery to the right of the *Resolution 1* placemark has a higher spatial resolution than the imagery to the left. Notice how individual features, such as trees, are easier to discern on the right. Zoom in and compare your ability to view the intricacies of the shoreline in the body of water on the right compared to the image on the left where features are much more generalized. Examine the *Res a* through *Res e* placemarks, zooming in as necessary.

Exploration 2.3: MULTIPLE CHOICE

1. Which of the following placemarks is located in the area with the lowest spatial resolution?

   A. Res a
   B. Res b
   C. Res c
   D. Res d
   E. Res e

You just had a taste of the way that different image sets overlapping can create varied impressions of a landscape. These differences can be caused by the fact that images have been captured by several sensors, have been taken at several times of the year, or were captured in different years. For example, double-click the *Imagery Sets* placemark. The varied imagery makes this region look dissimilar. Without zooming in, how many imagery sets can you identify in this view?

2. How many imagery sets does the view associated with the *Imagery Sets* placemark contain?

   A. 1
   B. 4
   C. 8
   D. 12
   E. >15

Go to the *Seasonal Differences* placemark. You will see a distinct contrast between the landscape on the right and left sides of the screen. Whenever you are considering the climate, agriculture, or vegetation of a scene, keep in mind the seasonal considerations. You can do this by using your common sense. For example, do the deciduous trees have leaves on them or not?

Additionally, you can usually find out the date the image was captured by moving your cursor over the landscape. The imagery date(s) will appear in the bottom left-hand corner of the screen. You need to be zoomed in sufficiently for the date to appear. Additionally, be wary of the fact that Google Earth™ will sometimes add color to imagery to help smooth the edges of image sets. This is quite apparent in one of the images associated with this placemark. Be careful not to let this added color bias your impression of a landscape.

3. Which of the following statements regarding the *Seasonal Differences* placemark is not supported?

   A. The imagery on the left was captured in the spring.
   B. The imagery on the right appears to have had color added.
   C. The imagery on the right was collected in January.
   D. The imagery on the left was captured in 2004.
   E. There is evidence of agricultural activity on both sides of the scene.

Seasonal variations in imagery can also impact your ability to interpret built environments. Urban forests can obstruct communities viewed from above. Go to the *Leaf-on* placemark and explore the image of a city captured in late spring. Then go to the *Leaf-off* placemark. The built environment is much more visible in the *Leaf-off* placemark. However, the location does not look as aesthetically appealing. Remember this factor as you interpret landscapes throughout this text.

Double-click the placemark *Imagery through Time*. You will zoom to an area near Birmingham, Alabama. The historical imagery function of Google Earth™ can be very helpful in highlighting change that takes place over time. In the Google Earth™ toolbar, click the Historical Imagery button. A time slider will appear in the upper left-hand corner of the image. Use this slider to see the landscape at various points in time. As you work through these exercises, keep in mind that every image is subject to change through time and seasonality. Turn off the Historical Imagery tool when you have completed the following questions.

4. Which of the following statements is not supported by the historical Google Earth™ imagery?
   A. High resolution imagery for the area is available in Google Earth™ dating back to 1978.
   B. Leaf-on and leaf-off imagery is available in the historical series.
   C. Buildings are constructed at the *Site A* placemark sometime between 1998 and 2002.
   D. *Site B* has forest cleared between 2005 and 2007.
   E. Roads are constructed at *Site C* between 2005 and 2006.

Exploration 2.3: SHORT ANSWER

1. What types of problems could arise when interpreting a scene that contains imagery from different years? Specifically, provide some examples of changes that can occur in both the physical and cultural landscape over time.

   _____
   _____
   _____
   _____
   _____
   _____
   _____

2. What are the most noticeable changes you can identify in the vicinity of the Imagery *through Time* placemark? How would you describe development in this region? Is it happening rapidly, moderately, or not at all? Support your answer with examples from the historical imagery.

   _____
   _____
   _____
   _____
   _____
   _____
   _____

Exploration 2.4: Interpretation

Finally, we will view a collection of locations that will help you develop some of the cognitive tools geographers use to evaluate and analyze the Earth from above. There are a number of "clues" present in any landscape that can help you understand and interpret the scene. These clues or elements include size, shape, texture, pattern, association, shadow, and site/situation.

Tone and/or color is one of the most distinguishing characteristics of a given feature in Google Earth™. Open the *Tone/color 1* placemark and you will see a large football stadium. The lush green of the pitch stands out. Another field might not be so well-maintained and would not be the deep, even green we see at Camp Nou, home to the FC Barcelona soccer team. We also see the distinctive red roofs of the region, the gray of the asphalt, and the numerous colors of automobiles. Now open *Tone/color 2* and view the coastal waters. The varied shades of blues and greens can indicate the presence of subsurface vegetation, algae, or coral structures. Here, the darker colors are related to deeper channels. *Tone/color 3* presents a homogenous scene of color.

Exploration 2.4: MULTIPLE CHOICE

1. The *Tone/color 3* placemark is centered in an area with relatively uniform color. This is a landscape dominated by

   A. cotton fields.
   B. forest.
   C. sand.
   D. snow and ice.
   E. water.

Size can help your interpretation in relative and absolute ways. Go to the *Size 1* placemark and you will see Boone Pickens Stadium at Oklahoma State University. We can obtain an absolute size from this image because we know that an American football field is 120 yards (about 110 meters) long if you include the end zones. You can apply this knowledge to measure the structure next to the stadium, Gallagher-Iba Arena. We also see opportunities for relative size evaluation. Automobiles can act as a comparative size reference.

Zoom to the *Size 2* placemark to see another really good example of an absolute size marker. The vast majority of railroad tracks in the US are a standardized gauge with the rails spread 4' 8.5" (1.48 meters) apart from one another. Go to the *Size 3* placemark.

2. If you wanted to determine the absolute size of the reservoir in the picture, what feature visible in the scene would be the best choice to use as your size reference point? In other words, which feature has a standard size?

   A. baseball field
   B. parking lot
   C. softball field
   D. tree crown size
   E. tennis court

Double-click the *Shape 1* placemark. You will fly to an industrial agriculture landscape. These are center-pivot irrigation schemes on the Great Plains. They utilize groundwater pumped to the surface and distributed by a sprinkler that rotates around a fixed pivot in the center of the field. Moving on to the *Shape 2* placemark, you will likely recognize this feature as it is named after its shape. The Pentagon in Washington, DC is one of the world's largest structures. The *Shape 3* placemark has some distinctive linear shapes. Explore this scene and identify the features.

3. The linear features associated with the *Shape 3* placemark are

    A. airport runways.
    B. interstate highways.
    C. streets in a housing development that has yet to be built.
    D. skyscrapers.
    E. railroad tracks.

The texture of a surface can be quite fine or very coarse. Texture can offer clues as to the types of vegetation on the ground or the degree of landscape homogeneity. Open *Texture 1* and view the dense forest. This is a relatively uniform landscape. *Texture 2*, on the other hand, has a significant degree of variation. You are viewing a transitional zone between steppe, woodlands, and forest from east to west. This changeover is a result of changing elevation and precipitation. Evaluate the *Texture 3* landscape.

4. Based upon your evaluation of the *Texture 3* landscape, identify the most accurate statement.

    A. This is a landscape with no variation in texture.
    B. This landscape has significant variation in texture as a result of agricultural activity.
    C. Texture varies in a regular and geometric pattern.
    D. The western side of this scene has much more variation in texture than the eastern side.
    E. This landscape has high texture variation because of its combination of vegetation and the built environment.

More often than not, regular patterns on the landscape suggest human involvement. For example, the *Pattern 1* placemark represents natural vegetation that is being intensively managed by humans. In this case, we are looking at one of the world's largest pecan orchards. The *Pattern 2* placemark highlights the concentric rings associated with surface mining. As this mining operation continues, this pattern grows in depth and width.

Now zoom to the *Pattern 3* placemark to view a pattern that is common across the western US. This is a planned suburb where the houses are all very similar in size and appearance. In the scene there are only two buildings that deviate significantly from the pattern. Zoom in and determine what these structures are.

5. The structures that deviate from the pattern in the *Pattern 3* scene are

    A. homes for the very wealthy.
    B. retail establishments.
    C. schools.
    D. factories.
    E. hospitals.

Association refers to the ability to identify or confirm the existence of a feature based on its relationship to other features. Double-click the *Association 1* placemark and you will find a collection of long yellow vehicles in a parking lot. These are school buses. Why would you have that many school buses in one location? Perhaps this is the bus yard for a large metropolitan school district? When we zoom out we see that it is not a large city, but there is a large industrial building adjacent to the buses. A little research reveals that this is Fort Valley,

Georgia, home of the Bluebird Bus Company factory.

Zooming to the *Association 2* placemark reveals strange groupings of what appear to be parts of airplanes. Let's zoom out and see if there is anything that would help confirm this. In fact, there are numerous intact airplanes and also an airport nearby that explains how this material has arrived. By the way, this is one of the US military's airplane "bone yards." They are usually found in desert areas, because the planes do not rust as quickly in the dry climate. Zoom to *Association 3* and use the associated features to help you determine what we are viewing.

6. The *Association 3* placemark is

    A. a railroad bridge.
    B. a dam.
    C. a combo rail and highway bridge.
    D. an aqueduct.
    E. a highway bridge.

Shadows can be helpful in providing an idea of a feature's height and shape as this information can sometimes be difficult to ascertain from a vertical perspective above the object. For example, the *Shadow 1* placemark illustrates Houston, Texas. Downtown is most distinguishable from its buildings' shadows. Excluding a few tall buildings due west of downtown, there is little vertical development of real estate outside the central business district.

The *Shadow 2* placemark illustrates a distinctive shadow associated with a cooling tower at one of France's largest nuclear power plants. Something is odd about this image though. Notice that the cooling tower is grainier than the rest of the image. Also, where is the shadow for the second cooling tower? This imagery may have been modified or distorted for security reasons. Now go to the *Shadow 3* placemark in Washington, DC.

7. Provide your best estimate as to what time of day this image was captured.

    A. 8 A.M.
    B. noon
    C. 2 P.M.
    D. 5 P.M.
    E. 8 P.M.

Site/situation refers to the geographic context of the feature, features, or landscape that you are interpreting. For example, go to the *Site/situation 1* placemark and you will see an airport with a number of planes and helicopters positioned on the tarmac. You also see a large number of modular structures around the specific location (the site). While site refers to the local circumstances, situation refers to the regional context. The situation of this image explains the aircraft and structures.

If we back out and turn on the *Borders and Labels* layer in the *Primary Database* you will see that we are in Afghanistan. This is a military base supporting the war efforts there. The situation of this image explains the aircraft and structures. *Site/situation 2* takes us to Iceland and the volcano that shut down air traffic over much of Europe in April of 2010. At the time this text was written, Google did not have imagery illustrating the eruption, but perhaps this has changed.

The moral of the story is: Google Earth™ imagery is not live. Sometimes you will find imagery that is very recent, while other times the imagery may be relatively old. *Site/situation 3* also underscores the necessity to be aware of contemporary contexts. We have zoomed to the grounds of a Presidential Palace. Why are thousands of people camped out around the site?

8. *Site/situation 3* illustrates

   A. the site of a recent earthquake.
   B. Earth Day.
   C. a massive demonstration against the government.
   D. the site of a recent tsunami.
   E. a street festival.

Exploration 2.4: SHORT ESSAY

1. Select one of the interpretation placemarks and describe the evidence of all of the interpretive elements or clues visible in that scene (tone, size, shape, etc.).

   _____
   _____
   _____
   _____
   _____
   _____
   _____

2. What two elements of interpretation (tone, size, shape, etc.) do you think are most helpful to you when trying to understand Google Earth™ imagery? Explain your answers and provide examples.

   _____
   _____
   _____
   _____
   _____
   _____
   _____

Encounter Human Geography

Name: _____

Date: _____

Chapter 3: Population

Understanding where the people of the world live and the changes that are taking place in the population structure provides a geodemographic background upon which to build our study of additional cultural attributes, such as political geography, development, and conflict. Population studies include elements such as the ratio of males to females, the percentage of the population in particular age groups, and the rate at which a population is growing. When viewed collectively, this geodemographic data can inform our understanding of contemporary and future challenges faced by countries and regions around the globe.

Download EncounterHG_ch03_Population.kmz from **www.mygeoscienceplace.com** *and open in Google Earth™.*

Exploration 3.1: Population

The total world population in 2011 is approaching seven billion people. However, these seven billion residents of Earth are not distributed across the face of the planet evenly. Countries range in population from more than one billion to less than one thousand. Open the *3. Population Issues* folder followed by the *Global Population Distribution* folder. Then, turn on the *Population by Country* folder. Manipulate the Earth to view the respective representations of each country's population. You can click on the flags of the countries to see specific values.

Exploration 3.1: MULTIPLE CHOICE

1.   Which of the following countries is not one of the world's five most populous?

   A. United States
   B. India
   C. Russia
   D. Brazil
   E. China

While the information from the *Population by Country* folder provides us with an understanding of population distribution on a country-by-country basis, it does not provide us with information on how people are distributed within these countries. Turn off the *Population by Country* folder and turn on the *World Population Density* folder. Be sure you have the *Borders and Labels* folder turned on.

2. Utilizing the *World Population Density* data, evaluate the following statements and identify which of the statements is not reflective of the data.

   A. China's population is disproportionately concentrated in the eastern half of the country.
   B. The most densely populated island in Indonesia is Sumatra.
   C. India's population is more uniformly distributed than China's.
   D. The population of Egypt is strongly associated with the Nile River.
   E. Australia's population centers are primarily located in coastal locations.

Turn off the *World Population Density* data. In the Layers pane, expand the *Gallery* folder in the *Primary Database*. Expand the *NASA* folder and then turn on the *Earth City Lights* layer. The bright areas in this data layer are the most urbanized, but not necessarily the most populated. For example, compare Western Europe with India.

3. Utilizing the *Earth City Lights* layer, evaluate the following statements. Identify the statement that is not supported.

   A. The lights of Pakistan are strongly associated with the Indus River.
   B. The lights of the US decrease in density west of the 97th meridian.
   C. The North Island of New Zealand displays more urbanization than the South Island.
   D. Nigeria has the highest amount of light pollution of any West African country.
   E. The state capital of Nevada is the most prominent light source in the state.

Turn off the *Earth City Lights* layer. From your analysis of the *World Population Density* data and the *Earth City Lights* layer you can see that there are extensive parts of the world with very low population density. Any number of factors can explain why people do not live in a given location. Oftentimes these explanations have an environmental root. Open the *Too What* folder and assess the physical environment of each placemark.

4. Pair the placemarks with the most likely environmental explanation for the low levels of human population near each placemark contained in the *Too What* folder.

   A. A = too wet, B = too cold, C = too high, D = too cold, E = too dry
   B. A = too cold, B = too dry, C = too wet, D = too high, E = too wet
   C. A = too dry, B = too hot, C = too cold, D = too dry, E = too wet
   D. A = too cold, B = too dry, C = too high, D = too wet, E = too dry
   E. A = too high, B = too wet, C = too dry, D = too cold, E = too cold

Exploration 3.1: SHORT ESSAY

1. Can you locate your city on the *Earth City Lights* layer? Is it a distinct entity or part of a larger agglomeration? What features in your vicinity can you identify by their light patterns?

_____

_____

_____

_____

_____

_____

_____

2. Describe and explain the stark difference between North Korea and South Korea in the *Earth City Lights* layer.

_____

_____

_____

_____

_____

_____

_____

Exploration 3.2: Increasing Population

While current population totals impact the contemporary policies of governments around the world, planning for the future is essential, as well. We can make projections based on current population statistics. Assess the current trends in population growth by turning on the *Population Growth Rate* folder.

Exploring 3.2: MULTIPLE CHOICE

1. Select the statement that best summarizes the regions of highest and lowest contemporary population growth rates.

    A. Africa = highest, Europe = lowest
    B. South Asia = highest, Africa = lowest
    C. East Asia = highest, North America = lowest
    D. South America = highest, Europe = lowest
    E. Europe = highest, Africa = lowest

The *doubling time* is the period of time required for a given population to double in number. It is obtained by dividing 70 by the percentage growth rate. For example, if a country has a growth rate of 3.5 percent, the population would double in approximately 20 years. Population growth rates for each country can be obtained by clicking on that country's flag.

2. Which of the following countries is most likely to double in population near the year 2030?

    A. Maldives
    B. Democratic Republic of the Congo
    C. South Africa
    D. Spain
    E. India

Turn off the *Population Growth Rate* folder and turn on *Total Fertility Rate* folder. Total fertility rate (TFR) is the average number of children a typical woman will give birth to in a given population. Based on what you have seen thus far, it should not be surprising that the highest TFRs are generally found in Africa.

3. What is the one exception of top 10 TFRs that is not in Africa or immediately adjacent to Africa?

    A. Yemen
    B. Mali
    C. Afghanistan
    D. Haiti
    E. Laos

Based on current populations, annual growth rates, and total fertility rates, you can now make some general projections as to the future population geography of the world.

4. Project the top five countries, in order, beginning with most populous for the year 2050.

    A. China, United States, Ethiopia, Pakistan, Bangladesh
    B. Indonesia, India, China, Pakistan, Vietnam
    C. Ethiopia, Congo, Nigeria, Uganda, Sudan
    D. India, China, United States, Indonesia, Pakistan
    E. United States, Russia, China, India, Nigeria

Exploration 3. 2: SHORT ESSAY

1. Why does Europe have a regional total fertility rate that is so different than much of the world? Use the internet to explore explanations for Europe's low fertility rates.

    _____
    _____
    _____
    _____
    _____
    _____
    _____

2. The country that is in the top 10 TFR that is outside of Africa faces unique challenges in its bid to lower birth rates. Discuss some of the reasons why TFR is high there. Use the internet to explore explanations for high fertility rates here.

_____
_____
_____
_____
_____
_____
_____

Exploration 3.3: The Demographic Transition

The demographic transition is a model that attempts to explain the transition through four or five stages of shifting birth and death rates. These rates can reflect a host of social and economic changes that occur in societies as they become more developed. The crude birth rate is calculated as the total number of live births per 1,000 persons in a given population. Open *The Demographic Transition* folder and then turn on the *Birth Rate* folder to compare birth rates around the world.

Exploration 3.3: MULTIPLE CHOICE

1. Order the following regions from highest to lowest according to birth rate: Europe, Latin America, North America, Southwest Asia/North Africa, Sub-Saharan Africa.

   A. Europe, North America, Southwest Asia/North Africa, Sub-Saharan Africa, Latin America
   B. North America, Europe, Sub-Saharan Africa, Southwest Asia/North Africa, Latin America
   C. Sub-Saharan Africa, Southwest Asia/North Africa, Latin America, North America, Europe
   D. Southwest Asia/North Africa, Latin America, Europe, North America, Sub-Saharan Africa
   E. Sub-Saharan Africa, Southwest Asia/North Africa, North America, Europe, Latin America

Likewise, the death rate indicates the total number of deaths in a population for every 1,000 persons. Turn off the *Birth Rate* folder and activate the *Death Rate* folder. Death rates reflect the current mortality situation in a country and can suggest conflict, economic hardships, health crises, and aging populations. The rate of natural increase that we viewed in the *Population Growth Rate* folder is a calculation based on the difference between birth rate and death rate.

2. Excluding Sub-Saharan Africa, which region has the highest death rates?

   A. South Asia
   B. Russian/Eastern Europe
   C. North America
   D. Southwest Asia/North Africa
   E. Latin America

3. Utilizing information from the birth rates, death rates, and the other population layers we have already viewed, order the countries placemarked with pink icons from earlier to later stages along the demographic transition. If necessary, use your textbook and/or the internet to see the stages of the demographic transition model.

    A. United States, Mexico, China, Pakistan, Bangladesh, Guatemala
    B. Honduras, Congo, Mexico, United States, China
    C. Iraq, Afghanistan, France, Canada, United States
    D. Indonesia, Mexico, China, Guatemala, Belgium, United States
    E. Pakistan, Guatemala, Indonesia, Mexico, China, United States

Tokyo, as well as much of Japan, is undergoing a period of significant population change. Turn off the *Death Rate* folder and double-click and turn on the *Tokyo Population*. Start the time slider animation located at the top of the screen by clicking the Play button on the time slider. Study the animation and evaluate the changes taking place in Tokyo. You can also view the data one frame at a time by clicking the Forward button on the right side of the time slider.

4. At what point did /does the majority of the city tip into declining population?

    A. 1990
    B. 2000
    C. 2010
    D. 2020
    E. 2030

Exploration 3.3: SHORT ESSAY

1. What are some future repercussions for Tokyo based on what you have seen in the animation? If necessary, search the internet for information as to what repercussions may occur as a result of high population densities, population decline, etc.

_____
_____
_____
_____
_____
_____
_____

Turn off the *Tokyo Population* folder and turn on the *Sex Ratio–65 Years and Older* folder. The sex ratio is the number of males for each female in a given society. This statistic can suggest the presence of sexual discrimination such as gender-specific abortions and infanticide.

2. Select a country at the high end and the low end of the range, respectively. Comment about possible explanations for these outliers.

_____
_____
_____
_____
_____
_____
_____

Exploration 3.4: Overpopulation

Another way to study population is to examine the proportion of a population in particular age brackets. For example, what percentage of a country's population is under the age of 15? This is a key statistic because young people require a lot of resources, but are not yet fully contributing to the economic output of a country. Turn off the _Demographic Transition_ folder and turn on the _%14 Years and Younger_ folder located inside the _Overpopulation_ folder.

Exploration 3.4: MULTIPLE CHOICE

1. On the basis of contemporary population growth and age structure, what region has the highest percentage of young dependents?

   A. East Asia
   B. Australia/New Zealand
   C. Northern Europe
   D. Sub-Saharan Africa
   E. South America

On the other end of the spectrum, a population that has a very high percentage of aged persons faces challenges as well. For example, many people are retired and are drawing on state pension systems without making continued contributions to the economic productivity of a country. Turn off the _%14 Years and Younger_ folder and turn on the _%65 Years and Older_ folder. A very different pattern emerges.

2. Utilizing both the _%14 Years and Younger_ and _%65 Years and Older_ folders, identify the statement that is most accurate.

   A. Chad's population is among the youngest, while Japan's population is among the oldest.
   B. Australia's population is among the youngest, while Samoa's population is among the oldest.
   C. Mali's population is among the youngest, while Madagascar's population is among the oldest.
   D. Honduras's population is among the youngest, while Canada's population is among the oldest.
   E. New Zealand's population is among the youngest, while Italy's population is among the oldest.

Many countries have instituted policies designed to limit population growth or encourage it. For example, financial incentives have been offered in France in recent years to encourage childbearing. In countries that have attempted to limit reproduction, couples have resorted to ultrasounds to identify the sex of the child in order to inform their decision on the baby's fate. In many cultures, a male child is preferred, and thus the ratio of males to females may be skewed. Close the *%14 Years and Younger* and *%65 Years and Older* folders and open the *Sex Ratio at Birth* folder to view the differences around the world.

3. Which of the following countries has the birth ratio that is least skewed toward males?

    A. China
    B. Azerbaijan
    C. India
    D. Liechtenstein
    E. Singapore

Turn off the *Sex Ratio at Birth* folder, turn on the *Sex Ratio 15–64* folder and Go to the United Arab Emirates. Then return to the folders (*%14 Years and Younger* and *%65 Years and Older*) that illustrate age proportions and examine the UAE rank in these two measures. Finally, view the *UAE 1* and *UAE 2* 360° city views.

4. What statement best explains the population characteristics of the UAE?

    A. Retirement center for the affluent.
    B. High demand for construction workers.
    C. Gender-specific abortion favoring females.
    D. Very young population because of high TFRs.
    E. Stage one of the demographic transition model.

Exploration 3.4: SHORT ESSAY

1. Regions with high percentages of young and/or aged persons face problems with high age-dependency ratios. This means that there is a significant portion of the population not involved in the labor force. Do you think it's more problematic to have a high percentage of aged persons or a high percentage of young persons?

    _____
    _____
    _____
    _____
    _____
    _____
    _____

Double-click on the *Global Projections of Drought* animation link. This documents changing drought regimes from 1870 to 2010. Read the embedded instructions for viewing layer animations. Analyze the trends and think about what the impacts will be on different regions of the world if the trends of the last century continue.

2. Do any regions appear to be particularly at risk for enhanced widespread drought? How
are Malthusian concerns impacted?

_____

_____

_____

_____

_____

_____

_____

_____

Name: _____

Date: _____

Chapter 4:
Human Health

The arena of human health is one of the more dynamic elements of human geography. Most corners of the world can, in theory, experience the range of human diseases. In reality, however, one is much more likely to die of malaria in Kenya and heart disease in the United States. These explorations aim to expose you to some of the rough geography of global health.

Download EncounterHG_ch04_HumanHealth.kmz from **www.mygeoscienceplace.com** *and open in Google Earth™.*

Exploration 4.1: Global Health Patterns

Open the *Human Health* folder and then open the *Global Health Patterns* folder. Turn on the *Life Expectancy at Birth* folder to view the global patterns for lifespan. Also, to find the life expectancy of an individual country, click on its flag. Because life expectancy measures the average number of years a newborn infant can expect to live at the current level of mortality, this statistic gives us a quick snapshot of the health of a country or region.

Exploration 4.1: MULTIPLE CHOICE

1. Identify the statement that is supported by the *Life Expectancy at Birth* folder.

    A. Life expectancies tend to be greater in far Northern Africa than in the rest of Africa.
    B. Life expectancies generally increase from west to east across Europe.
    C. The United States has one of the highest life expectancies in the world.
    D. All of the 20 lowest life expectancy countries are found on the African continent.
    E. There are no southern hemisphere countries with life expectancies that exceed 80 years.

Because hospitals and trained healthcare providers can profoundly influence the likelihood that a child will survive his or her first year of life, infant mortality rate can provide a measure reflective of a country's healthcare infrastructure. Turn off the *Life Expectancy at Birth* folder and turn on the *Infant Mortality Rate* folder and assess the global patterns.

2. Identify the statement that is supported by the *Infant Mortality Rate* folder.

    A. In the United States, approximately one in six infants die.

    B. Infant mortality rates in all East Asian countries tend to be among the best 10 percent of all countries.

    C. In 2008, there were more than ten countries in the world where at least one of every ten infants died.

    D. In 2008, the lowest infant mortality rate in the world was found in Europe.

    E. In 2008, the highest infant mortality rate in South America was found in Venezuela.

HIV/AIDS has killed more than 25 million people since it was first recognized in the early 1980s. HIV/AIDS has left its mark around the world in both developed and developing regions. Turn off the *Life Expectancy at Birth* folder and turn on the *People Living with HIV/AIDS* folder.

3. Identify the statement that is supported by the *People Living with HIV/AIDS* folder.

    A. In 2008, Russia, China, the United States, and Brazil were top ten countries in terms of persons living with HIV/AIDS.

    B. There were less than 300,000 persons with HIV/AIDS living in Europe in 2008.

    C. The data suggests that the hot spot for HIV/AIDS infections is North Africa.

    D. Ethiopia, Kenya, and Tanzania each have more than two million infected persons.

    E. In 2008, South Africa, India, and Nigeria had the greatest number of persons living with HIV/AIDS.

While one can get a sense of the parts of the world that are most impacted by HIV/AIDS from the *People Living with HIV/AIDS* folder, the proportional impact of HIV/AIDS on the countries of the world is unclear. Keep the *People Living with HIV/AIDS* folder active and turn on the two remaining HIV/AIDS datasets: *HIV/AIDS Annual Deaths* and *HIV/AIDS Adult Prevalence*. Explore the varying impacts of HIV/AIDS on countries and regions around the world.

4. Identify the statement that is supported by the three HIV/AIDS folders.

    A. The greatest concentration of the HIV/AIDS pandemic lies within the southern tip of Africa.

    B. Approximately 6 percent of American adults have HIV/AIDS.

    C. India is home to both the highest prevalence rate and the highest number of persons infected with HIV/AIDS.

    D. The highest numbers of deaths due to HIV/AIDS in 2008 occurred near the line of 40° west longitude.

    E. Romania has felt the impact of HIV/AIDS more than any other eastern European country.

Exploration 4.1: SHORT ESSAY

1. Google™ the HIV/AIDS pandemic and provide some insight into why there is such a regional distinction in terms of the most severely impacted region of the world. What are some ways that the situation can be improved in this region in the future?

    _____

    _____

    _____

    _____

    _____

    _____

    _____

2. Were you surprised to see the United States ranking in the *Life Expectancy at Birth* folder? Describe some factors that could explain the higher ranked countries.

    _____

    _____

    _____

    _____

    _____

    _____

    _____

Exploration 4.2: Developing World Health Threats

Not surprisingly, health threats in the developed and developing world are often radically different. Many situations that are taken for granted in the more developed countries are a daily struggle in the less developed countries. Perhaps the best example of this is centered on the issue of clean water. Access to potable water can be particularly problematic in the rural areas of developing countries. Turn on the *Rural Access to an Improved Water Source* folder.

Exploration 4.2: MULTIPLE CHOICE

1. Identify the statement that is supported by the *Rural Access to an Improved Water Source* folder.

    A. Mongolia has higher rates of access to improved water sources than adjacent countries.
    B. Despite limited access to potable water throughout much of Africa, Mozambique is a global leader in clean water access in rural areas.
    C. Cambodia has higher rates of access to improved water sources than adjacent countries.
    D. Excluding Africa, Central Asia has one of the lower regional rates of access to improved water sources.
    E. South America generally has higher regional rates of access to improved water sources than North America.

Likewise, access to improved sanitation is an important barometer of development and human health. When people live with inadequate sanitation, like open sewers, they are much more susceptible to diseases such as Cholera. Close the *Rural Access to an Improved Water Source* and open the *Access to Improved Sanitation* folder.

2. Identify the most dramatic difference between adjacent countries in terms of *Access to Improved Sanitation*.

    A. Moldova-Ukraine
    B. Nicaragua-Costa Rica
    C. United States-Mexico
    D. Papua New Guinea-Indonesia
    E. Libya-Chad

When one becomes ill, it is important that a physician can attend to him or her. In many countries around the world, the number of physicians available per person is much lower than what is commonly found in the United States. Close the *Access to Improved Sanitation* folder and open the *Physicians per 100,000 People* folder.

3. Identify the statement that is supported by the *Physicians per 100,000 People* folder.

    A. Communist and formerly Communist countries generally have lower physician per capita figures than capitalist countries of North America and Western Europe.
    B. There is a general north-south global divide for number of physicians per 100,000 persons.
    C. The southern tip of Africa is the most underserved region in terms of physicians per 100,000 persons.
    D. Eurasia trails Latin America in terms of physician access.
    E. Paraguay has the highest level of access to physicians.

The Centers for Disease Control and Prevention (CDC), based in Atlanta, is a global leader in the protection of public health and safety. One way it serves this mission is through the dissemination of a wide variety of information. Open the link associated with the *Centers for Disease Control and Prevention*. You will see a typical disease information page. In this case, malaria information is provided. As a reminder, it is usually best practice to open the page outside Google Earth™ by clicking the "Open in Google Chrome" button.

4. Based on the CDC information, identify the most accurate statement regarding malaria.

    A. Malaria is present throughout the South American country of Chile.
    B. Malaria has been eliminated from Vietnam.
    C. Unfortunately, anti-malarial drugs have yet to be developed.
    D. The most vulnerable group to contracting malaria is young children.
    E. *Anopheles* refers to an experimental malarial drug.

Exploration 4.2: SHORT ESSAY

1. In your opinion, which of the folders included in this exploration provides the clearest distinction between developed and developing regions? Explain your answer.

_____

_____

_____

_____

_____

_____

_____

2. Use the CDC Web site to explore another disease that is more common in the developing world. Provide some background information on the disease and describe its geographic distribution.

_____

_____

_____

_____

_____

_____

_____

Exploration 4.3: Developed World Health Threats

While people generally live longer in the developed world, there are health threats that disproportionately affect this segment of the global population. For example, our highly industrialized economies that have contributed to such prosperity have their own set of drawbacks. For example, polluted air and water sources lead to higher cancer rates than those found in the developing world. The Environmental Protection Agency (EPA) monitors air and water pollution in the United States. Turn on and open the _EPA Air Emission Sources_ folder. You will find a collection of folders inside that illustrate different types of air emissions. Explore the individual folders to help develop an understanding of the information.

Exploration 4.3: MULTIPLE CHOICE

1. Using the information contained in the _EPA Air Emission Sources_ folder, identify the most accurate statement.
   A. Oil and production emissions are clustered in the Pacific Northwest.
   B. The Dakotas have a disproportionately high number of electric generating units.
   C. Petroleum refineries are disproportionally located near coastlines.
   D. The western half of the United States has disproportionately more chemical manufacturers than the eastern half.
   E. The pulp and paper industry is clustered near the forests of the Rocky Mountains.

You can tilt the display of _EPA Air Emission Sources_ to find big emitters as the height of each point corresponds to the total emissions from the facility. If you need help remembering how to tilt the 3D Viewer, consult the Google Earth™ User Guide. You can even click on each placemark to see the emissions report broken down by pollutant for each site.

2. Study the biggest emitters by type and identify the statement that is supported.

   A. In 2005, the greater Los Angeles area did not have any facilities that exceeded the 2,000 ton mark for carbon monoxide emissions.
   B. Sulfur dioxide is the most highly emitted pollutant in the Indianapolis metropolitan area.
   C. The largest emitters in Georgia are releasing lead as the primary pollutant.
   D. Pulp and paper industries in the state of Washington emit more volatile organic compounds than any other pollutant.
   E. The top emitters in Alabama are associated with chemical manufacturing.

Turn off the *EPA Air Emission Sources* and then go the *Times Beach* placemark. Evaluate the site around the placemark and then use the Web to determine what transpired here.

3. Identify the statement that is not supported by your research and assessment of the *Times Beach* placemark.

   A. People are strictly forbidden to enter the former town site.
   B. The 1990 historic imagery shows that streets and homes were still present at that time.
   C. The impacted area has now been converted to a state park.
   D. The Environmental Protection Agency bought the entire town.
   E. The chemical exposure that occurred at Times Beach is one of the largest of its kind in the history of the United States.

Now perform the same action for the *Love Canal* placemark.

4. Identify the statement that is not supported by your research and assessment of the *Love Canal* placemark.

   A. Since 1995, there have been new structures built on the site.
   B. Love Canal is a neighborhood located in Niagara Falls, New York.
   C. A school was constructed on this toxic waste site in the 1950s.
   D. Love Canal was instrumental in the development of the Superfund program.
   E. No adverse health impacts to residents of the area were ever proven.

Exploration 4.3: SHORT ESSAY

1. Open the *Air Quality from AIRNow* folder and explore the data for yesterday, today, and tomorrow. Find the location that is nearest to you and report on the conditions. Name a few locations reporting problems with air quality.

   _____

   _____

   _____

   _____

   _____

   _____

   _____

2. Turn on and open the *EPA Air Emission Sources* folder again and identify two of the polluters nearest you. Were you aware of these facilities? What are the primary pollutants? Use the internet to find out some of the adverse health effects associated with the pollutants you listed.

   _____

   _____

   _____

   _____

   _____

   _____

   _____

Exploration 4.4: The Future of Health

One can look to the past for stark examples of the power of disease to impact the trajectories of human societies. Perhaps the most prominent example is the Bubonic Plague sweeping across Europe in the twelfth century. Perhaps one-third to one-half of the population of Europe died and a dramatic social and economic transformation followed.

While Bubonic Plague rarely makes the news today, there are still several thousand cases that occur annually. Diseases like plague and anthrax do not pose the same level of threat in our everyday environment as in the past, but the weaponization of these diseases is a concern for the future. The World Health Organization estimates that a deliberate attack with aerosolized inhaled *Y. pestis* bacilli (pneumonic plague) over a city of five million would affect as many 150,000 persons with 36,000 fatalities. As devastating as that sounds, it pales in comparison to the European outbreak of plague in the twelfth century.

Open the *Future of Health* folder and turn on and open the *Bubonic Plague in Europe* folder. If you turn off the *Borders and Labels* folder it will help unclutter the map. Begin the animation by clicking the play button in the timeline located at the top of the screen. It will likely take a few repeats for all of the images to load. Remember you can adjust the transparency of the layer by using the slider at the bottom of the Places pane. You may need to do this to locate cities not located on the historic map.

Exploration 4.4: MULTIPLE CHOICE

1. Evaluate the progression of the Bubonic Plague in Europe using the animated map. Identify the statement that is supported.

   A. Northeastern Europe was the first part of the continent impacted.
   B. Vienna was impacted before Barcelona.
   C. The disease generally progressed from north to south.
   D. The map suggests that all of Europe was affected by the Black Death.
   E. London was likely affected before Budapest.

Close and turn off *The Bubonic Plague in Europe* folder and turn on the *Immunization Rate for DPT* folder. Diphtheria (DPT) is an upper respiratory tract disease that was very common historically. However, a vaccine for the disease has been very effective at reducing transmission. For example, there have been less than 100 cases in the United States in the last 25 years. Where the disease has not been eradicated, it is still quite dangerous with fatality rates from 5 to 20 percent. In the 1990s there was a great increase in diphtheria in one world region, becoming one of the world's most resurgent diseases. Google™ diphtheria using outside resources and evaluate the distribution of immunization in Google Earth™.

2. Which of the following statements is supported by your outside research and the *Immunization Rate for DPT* folder?

   A. In 2005, the highest rates for DPT immunization were found in southern Africa.
   B. The break-up of the Soviet Union led to a surge of diptheria in Russia and former Soviet states.
   C. In 2005, India had a higher diphtheria immunization rate than China or Brazil.
   D. Diphtheria is spread primarily through mosquito bites.
   E. Diphtheria has been completely eradicated from Southeast Asia as a result of the region's 100 percent immunization record.

Turn off the *Immunization Rate for DPT* folder and turn on the *Alcohol Consumption per Person* folder. This folder effectively highlights the wide range in alcohol consumption found globally. These stark differences can be the result of economic gaps and/or cultural norms. One thing that is not reflected in this map is the ongoing global shift toward greater consumption of alcoholic beverages. Needless to say, this trend will lead to accumulating adverse health impacts.

3. Use the *Alcohol Consumption per Person* folder along with your knowledge of cultural norms to select the statement that is most likely to be accurate.

   A. Alcohol consumption per capita in the United Arab Emirates is very low because of poverty.
   B. Alcohol consumption in Europe leads the globe from a regional perspective.
   C. From a regional perspective, Latin America has the lowest per capita consumption rates.
   D. Alcohol consumption in Iran is higher than most countries of the world.
   E. Consumption of alcohol in North Korea is higher than in South Korea.

Alcohol consumption is but one reflection of the world's changing lifestyles. Another one that arguably creates a much larger impact on not only health, but the environment, is the trend toward increased consumption of meat. Turn off the *Alcohol Consumption per Person* folder and expand and turn on the *Greenpeace–Land Use* folder. This folder highlights changes occurring in the state of Mato Grosso in Brazil. Explore the placemarks and read the information provided by Greenpeace by clicking each link.

4. Which of the following statements is not supported by the *Greenpeace–Land Use* folder?

   F. Cattle ranching is expanding in Brazil.

   G. Deforestation is clearly identifiable in the surrounding states of Rondônia, Para, and Amazonas.

   H. Near the *Amazon Biome Strip* placemark, one can see the last remaining strips of rainforest in Mato Grosso.

   I. Beyond the western extent of cattle ranching in Mato Grosso, one enters the area of soya farming.

   J. Forest destruction for cattle ranching is damaging the springs of the Xingu River.

Exploration 4.3: SHORT ESSAY

1. Discuss how the changing political/economic geography of the twelfth century likely contributed to the occurrence of the Black Death in Europe.

_____

_____

_____

_____

_____

_____

2. Alcohol and meat consumption are two variables that reflect the changing lifestyles of people around the world. As the middle class grows and people become more affluent, we can expect to see more lifestyle changes. Think of another measure that would reflect these changes. Explain the patterns you would see on the Earth if you mapped it in the same way we have looked at other variables on Google Earth™.

_____

_____

_____

_____

_____

_____

Encounter Human Geography

Name: _____

Date: _____

Chapter 5:
Migration

Migration, or the movement of people from one location to another, has been an essential part of human behavior for all time. Humans continually seek better places to live for a variety of reasons including issues related to safety, opportunity, climate, or freedom. People migrate from one country to another or within countries. Along the way they are likely to face any number of obstacles. The following explorations aim to provide you with insight into the always dynamic realm of human migration.

Download EncounterHG_ch05_Migration.kmz from **www.mygeoscienceplace.com** *and open in Google Earth™.*

Exploration 5.1: Reasons for Migrating

Many people around the world are forced to migrate due to significant push factors. When large numbers of migrants move to one particular area, the resources of the receiving area can be overwhelmed. In these cases, the work of international agencies such as the United Nations High Commissioner for Refugees (UNHCR) is essential as it is the international agency charged with the protection of refugees around the world. In 2009, the UNHCR identified more than ten million refugees worldwide. Many of these refugees end up in sprawling refugee camps. Open the *Reasons for Migrating* folder followed by the *Refugee Camp in High Resolution* folder. Check to make sure the sound on your computer is turned on and play the refugee camp tour. If you have a slower connection, the tour will work better the second time you play it.

Exploration 5.1: MULTIPLE CHOICE

1. After watching the refugee camp tour, identify the statement that is not supported.

   A. Refugee camps can consist of hundreds of shelters.
   B. Refugee camps are found in a variety of physical terrain, from forests to deserts.
   C. The refugee camps in the tour are primarily located in rural areas.
   D. The refugee camps are protected by walls and security forces with tanks and artillery.
   E. Many of the refugees are living in non-permanent structures, such as tents.

In the *UNHCR* folder, there are placemarks for the seven camps you visited on the tour. Click the respective links to view the specific data for each camp.

2. After viewing the attribute data for each refugee camp, identify the statement that is not supported.

   A. The camps all have populations of 15,000 persons or more.
   B. At least one of the featured camps has been in place for more than 25 years.
   C. At Gihembe, the average quantity of water available per person is approximately the same amount that a typical 3.5 gallon (13 liter) per flush toilet uses in a single flush.
   D. In general, many people have to share access to usable water taps.
   E. The featured camps are all located in the Southern Hemisphere.

Follow the *one more click* link available on any of the refugee camp data pages to link to the homepage of UNHCR by selecting the blue EN box. A variety of data is available from UNHCR, including information on the different groups of people that UNHCR attempts to help.

3. Explore the information and maps on the UNHCR Web site related to where the organization works. Under the Where We work heading, find out which country in Europe hosts the largest refugee population.
   A. France
   B. Germany
   C. Finland
   D. Serbia
   E. Italy

One key distinction in terminology is the difference between a refugee and an internally displaced person, or IDP. A refugee is a person who has crossed an international boundary because of persecution in their own country. This persecution can take many forms, such as religious, ethnic, racial, or political. An internally displaced person, on the other hand, may face the same persecution, but remains in their country of origin, though not in their home. Turn on the *Internally Displaced Persons* folder.

4. In 2006, which of the following countries had the highest population of internally displaced persons?
   A. Colombia
   B. United States
   C. Guatemala
   D. India
   E. Uganda

Exploration 5.1 SHORT ESSAY

1. The number of persons per usable water tap was one of the figures listed for the refugee camps. Why is this statistic relevant?

    _____
    _____
    _____
    _____
    _____
    _____
    _____
    _____

2. Explain the factors that led to the excessively high level of internally displaced persons for the leading country in 2006. Provide an update on the current situation in this country.

    _____
    _____
    _____
    _____
    _____
    _____
    _____
    _____

Exploration 5.2: Distribution of Migrants

Turn off the *Reasons for Migrating* folder and open the *Distribution of Migrants* folder. The first folder illustrates gross domestic product per person (*GDP per Capita*). This is a general measure of the affluence of each country. As you can see, the range from highest to lowest is dramatic. One of the strongest pull factors in migration is the perceived opportunity to better one's self economically. Not surprisingly, there is a strong relationship between the next folder, *Net Migration Rate*, and the measure of affluence. People tend to migrate to the more affluent countries.

Exploration 5.2: MULTIPLE CHOICE

1. Select the statement that is not supported by the data from the *Net Migration Rate* folder.

    A. Latin America is generally a region of out-migration.
    B. Afghanistan is the leading country of out-migration in 2008.
    C. Western Europe is a receiver of migrants while Eastern Europe is a source of migrants.
    D. Cuba is an out-migration country in 2008.
    E. North America is generally a region of in-migration.

2. Upon comparing and contrasting the *Net Migration* folder and the *GDP per Capita* folder, which of the following statements is not supported.

   A. Not all countries in the upper half of GDP are net in-migration countries.
   B. Liberia has the strongest logical connection between affluence and migration.
   C. Not all countries in the lower half of GDP are net out-migration countries.
   D. Qatar's relationship between affluence and migration is counterintuitive.
   E. Costa Rica has a relatively high GDP for Central America and also a relatively high in-migration rate for Central America.

Turn off the *Net Migration* and the *GDP per Capita* folders. The United States has an interesting history of immigration (in-migration). The dominant regions of origin for these migrants have varied widely over the last 200 years. Read your text and any outside resources to gain an understanding of the general evolution of US immigration and then open the *Countries of Birth* folder. This folder contains five folders (*Year V–Year Z*) of countries representing the top five source countries at different years in American history (1850, 1900, 1930, 1970, and 2000). The countries within each group represent the first through fifth highest source countries for the aforementioned years. For example, V1 was the top source country in a given year and V5 was the fifth highest source country in that same year.

3. What folders represent US immigration in 1850 and 2000?

   A. 1850 = Z, 2000 = W
   B. 1850 = X, 2000 = Z
   C. 1850 = V, 2000 = Y
   D. 1850 = W, 2000 = V
   E. 1850 = W, 2000 = Y

Migration often occurs from one specific location to another because of an established ethnic or national population at the destination location. The established population provides a social, cultural, and economic network that eases the transition to the new destination for immigrants. As a result, unique districts that reflect this process can emerge within urban areas. To see an example of this type of migration go to the *Migration Related District* placemark.

4. Examine the street scene of the *Migration Related District* placemark. What migration related term best applies to this location?

   A. Intraregional migration
   B. Chain migration
   C. Internal migration
   D. Push factor
   E. Counterurbanization

Exploration 5.2 SHORT ESSAY

1. Explain the clear discrepancies between Eastern and Western Europe that can be seen in the *GDP per Capita* and *Net Migration Rate* folders.

_____
_____
_____
_____
_____
_____
_____

2. Explain the process that was highlighted with the *Migration Related District* placemark. How or why does it happen and grow?

_____
_____
_____
_____
_____
_____
_____

Exploration 5.3: Obstacles to Migration

Open the *Obstacles to Migration* folder and then explore the *View from Ellis Island*. This is a perspective that many immigrants to the United States would have experienced if entering the country from the late nineteenth to the mid-twentieth century. At Ellis Island, one obstacle immigrants faced was the medical examination. Many unhealthy persons were denied entrance to the country. While Ellis Island is synonymous with New York City, the majority of the island is not located in New York.

Exploration 5.3: MULTIPLE CHOICE

1. Use Google Earth to determine what state claims the majority of Ellis Island.

   A. Connecticut
   B. Delaware
   C. New Jersey
   D. New York
   E. Pennsylvania

While Ellis Island most often comes to the forefront of discussions on historic US immigration stations, there was another heavily utilized station. Angel Island was the point through which many immigrants passed if arriving from sources to the east. Click the *Angel Island, CA* placemark to see the site. There are several Web sites learn more about the island in the *Angel Island Websites* folder.

2. Utilize Google Earth imagery and outside resources to learn more about this key location in the immigration history of the United States. Select the statement that is not supported by your research.

   A. Angel Island is now a California State Park.
   B. Angel Island is located in San Francisco Bay.
   C. The welcome to incoming Asians on Angel Island was generally more receptive than that given to Europeans at Ellis Island.
   D. Australians, New Zealanders, and Russians were among the immigrants for the immigration station at Angel Island.
   E. Chinese immigrants were singled out upon arrival at Angel Island.

Canada also has historically been a destination for migrants from around the world. Grosse Île is home to a monument that commemorates the role the island played from 1832–1937 as a port of entry into Quebec. More specifically, Grosse Île was established as a quarantine station for incoming migrants. In 1847, the island was overwhelmed with an epidemic that left thousands dead. Research the site using outside resources and the *Photos* layer.

3. The island of Grosse Île was hit by a pandemic in 1847 that left thousands of migrants dead. What was the migrant group that was most affected and what was the disease?

   A. Potawatomi Indians, Smallpox
   B. African, Bubonic Plague
   C. Quebecois, Cholera
   D. German, Influenza
   E. Irish, Typhus

Open the *Border Crossings* folder and look at US–Mexico crossing at Tijuana, the US–Canada crossing at Detroit and Windsor, and the section of the French/Swiss border. Notice the facilities for processing border traffic along each border as well as evidence of a fortified border or an open border.

4. Based on the *Border Crossings* placemarks identify the statement that is most correct.

   A. The French-Swiss border is characterized by multiple walls between the two countries.
   B. The Tijuana crossing has the most freight (semi tractor trailers) at the border.
   C. The line of traffic to cross the Swiss border is longest.
   D. The Detroit crossing has the largest footprint of any of the crossing stations.
   E. Passenger automobiles and semi tractor trailers are not separated at any of the US crossings.

Exploration 5.3 SHORT ESSAY

1. Ellis Island has a unique geopolitical history. Research the issue of state sovereignty on Ellis Island and provide a brief synopsis.

_____
_____
_____
_____
_____
_____
_____

2. One can see the remarkable build-up of traffic along the US–Mexico border at the Tijuana crossing. Is this normal? View the historical imagery of the Tijuana border crossing and evaluate the traffic crossing the border through time. Is there heavier traffic now than in the past? Is there a pattern of heavier traffic from the US or Mexican side of the border?

_____
_____
_____
_____
_____
_____
_____

Exploration 5.4: Internal Migration

Exploration 5.4: MULTIPLE CHOICE

Ten placemarked states are contained within the _Net Domestic Migration by State_ folder that is contained within the _Internal Migration_ folder.

1. Based on what you have learned about internal migration patterns in the United States from your course and your text, identify the list of placemarked states that have most likely been net in-migration states versus net out-migration states in the last two decades.

   A. In-migration = Arizona, Georgia, Nevada; Out-migration = Illinois, Massachusetts, New York

   B. In-migration = California, Texas, Florida; Out-migration = Michigan, Ohio, Pennsylvania

   C. In-migration = Georgia, Nevada, New York; Out-migration = California, Georgia, Massachusetts

   D. In-migration = California, Florida, New York; Out-migration = Arizona, Florida, Georgia

   E. In-migration = Arizona, California, Nevada; Out-migration = Alabama, Georgia, Florida

Turn off the _Net Domestic Migration by State_ folder and turn on and expand the _Migration Focus Counties_ folder. The counties contained in this folder have been national leaders in the last decade in terms of net migration percentage change. Several of the counties have been net gainers while several have been net losers. Zoom in to the counties and survey the landscape to help inform you as to the factors that are at play for each county. Perhaps it is a county dependent on agriculture or a deindustrializing area. For example, is the county gaining migrants as a result of suburban sprawl or in a declining urban core? Using the historical imagery can provide clues, as well. For your information, four of the counties are net loss counties and two are net gain counties.

2. Select the assessment that is supported by the imagery and any additional outside research you perform.

   A. Loudon County is likely a net loss county due to increasing agricultural efficiency, while Seward County is a net gain county because of a rapidly growing metropolitan area.

   B. Arlington County is likely a net loss county because of deindustrialization while Genesee County is likely a net gain county as a result of a booming agricultural-based economy.

   C. Chattahoochee County is likely a net gain county because of its proximity to Columbus, Georgia, while Genesee County is likely a net loss county because of deindustrialization.

   D. Lincoln County is likely a net loss county due to increasing agricultural efficiency, while Arlington County is likely a net gainer as a result of expanding Washington, DC suburbs.

   E. Seward County is likely a net loss county because of increasing agricultural efficiency, while Loudon County is likely a net gainer because of expanding Washington, DC suburbs.

3. Which of the net growth counties has the most visible signs of recent conversion of agricultural land to suburban development? Remember the historical imagery capabilities of Google Earth™.

   A. Arlington County, Virginia
   B. Chattahoochee County, Georgia
   C. Genesee County, Michigan
   D. Lincoln County, South Dakota
   E. Seward County, Kansas

Turn off the *Migration Focus Counties* folders and go to the *Brasilia* placemark. Brasilia was planned, created, and developed in the late 1950s and became the national capital in 1960. Its location was selected to help draw Brazilians into the interior of the country and away from the heavily populated southeast. Today, the metropolitan area is home to more 3.5 million persons. This represents a prime example of recent internal migration. Many of the workers who helped build the futuristic city decided to stay when the job was complete. They sent for their families, and patterns of chain migration were established. As many of the migrants were simply seeking the opportunity to advance themselves economically, they came with limited resources and began to establish squatter communities in and around the city. This clashed with the government's desire for Brasilia to be a clean and tidy vision of Brazil's urban future.

4. Explore Brasilia, using the aerial imagery, the historic imagery, and the *Photos* layer. Identify the statement that is not supported by your exploration of Brasilia.

    A. The core of the city appears as an airplane when viewed from above.
    B. The squatter settlements are clustered along the central spine of the master-planned development.
    C. The most densely populated areas of the city are not found within the master planned development.
    D. The central spine of the city contains a number of large monuments and structures.
    E. Large boulevards and highways help define the symmetrical shape of the central city.

Exploration 5.4: SHORT ESSAY

1. View the *Chevy in the Hole* polygon. Google™ "Chevy in the Hole" to determine the nature of this site. How is it representative of Genesee County's recent migration history?

    _____
    _____
    _____
    _____
    _____
    _____
    _____

2. View the *Raspberry Falls Golf and Hunt Club* placemark. How is this location representative of Loudoun County's recent migration history? You may find it helpful to research the term "bedroom community."

    _____
    _____
    _____
    _____
    _____
    _____

Name: _____

Date: _____

Chapter 6:
Folk and Popular Culture

Exploration 6.1: Folk and Popular Places

People around the world live life in very different ways. Some of your own traditions and habits have likely been passed down for generations, while others may have been picked up last week. In the United States, your way of life is likely to incorporate a high degree of popular culture. Other parts of the world still employ a higher degree of traditional or folk cultures. Because of globalization, however, it is increasingly difficult to find unadulterated folk culture. Synthesis of cultures is becoming the norm. The following explorations expose you to some of the varied elements of folk and popular culture.

Download EncounterHG_ch06_Folk_and_Popular_Culture.kmz from **www.mygeoscienceplace.com** *and open in Google Earth™.*

Exploration 6.1: MULTIPLE CHOICE

Open the *Folk and Popular Places* folder and then open the *Folk and Popular Music* folder. There are four genre folders that contain placemarks associated with different popular and folk music styles. Explore the locations and use your text and outside resources to verify what type of music is associated with each genre folder.

1. Identify the statement that is <u>not</u> supported by the genre placemarks and your outside research.

    A. Bands like Nirvana and Pearl Jam are associated with the *Genre 1* folder.
    B. The Grand Ole Opry would be a logical addition to the *Genre 2* folder.
    C. Woody Guthrie is associated with the *Genre 4* folder.
    D. *Genre 1* and *Genre 2* are better examples of popular music while *Genre 3* and *Genre 4* are better examples of folk music.
    E. The "Grunge Rock" associated with the *Genre 1* folder was the most recently developed genre of the four listed in Google Earth™.

Tattooing represents a distinctive element of cultural geography. Tattooing has been a part of the folk culture of Polynesian people for more than 2,000 years. In the last 20 years, the popularity of tattooing has exploded in the western world. View the placemarked locations in the *Folk or Popular Tatts* folder. To see the storefront of *Miami Ink*, zoom in to Street View.

2. Identify the statement that is supported by the *Folk or Popular Tatts* placemarks and/or your outside research.
    A. At *Tieke Marae*, one would potentially acquire a bicep tattoo of Yosemite Sam with text that says "Back Off!"
    B. Maori tattoos are exclusively a part of popular culture.
    C. The tattoos at the *Tieke Marae* location would most likely be chiseled into the facial region with a bone chisel.
    D. The tattoo is a good example of popular culture diffusing into folk culture.
    E. *Miami Ink* specializes in folk-style tattooing.

In the *Folk or Popular Sport* folder you will find three examples of sports from around North America. Examine these placemarks and do any research needed to gain an understanding of what sports are associated with each location.

3. Identify the statement that is supported by the *Folk or Popular Sport* placemarks and/or your outside research.

    A. Baseball originated at the Springfield, Massachusetts YMCA.
    B. Beaver, Oklahoma, claims to be the original home of soccer.
    C. Of the sports highlighted in this folder, basketball's origin has the strongest folk roots.
    D. Cow chip tossing is a popular sport associated with urban locations.
    E. The sport associated with the Air Canada Centre is a popular sport with a folk origin.

Look at the placemarked locations included in the *Folk or Popular Lifeway* folder. Identify clues that suggest elements of folk and popular culture. In this case, think of culture as a way of life. Think about influences such as the level of technology at each location and any special circumstances in terms of ethnic or religious groups. You may need to use outside resources to learn more about the ethnicities and religions of each area.

4. Identify the most likely place for elements of folk culture in the locations as they are shown.

    A. Chenonceaux, France
    B. Kalona, Iowa
    C. Kandahar, Afghanistan
    D. Libreville, Gabon
    E. Ulaanbaatar, Mongolia

Exploration 6.1 SHORT ESSAY

1. When you think about the significance of Springfield, Massachusetts, as the point of origin for a popular sport, do you see any geographic evidence that this site has influenced the contemporary geographic distribution of this sport? For example, is it still most popular in its hearth area or has the sport expanded beyond its initial hearth?

    _____
    _____
    _____
    _____
    _____
    _____

2. Revisit the *Folk or Popular Lifeway* folder. Briefly describe one element of popular culture for each location. For example, Chenonceaux is an older location of settlement, but the highly manicured grounds and the elaborate chateaux do not suggest a simple folk culture.

_____
_____
_____
_____
_____
_____

Exploration 6.2: Folk Clustering

In a few locations, one can still see significant evidence of folk lifeways at work in our modern world. This includes examples in the United States. Open the *Folk Culture in the US* folder within the *Folk Clustering* folder and examine the two placemarks to identify the group associated with this example of folk culture.

Exploration 6.2: MULTIPLE CHOICE

1. Select the folk group represented in the *Folk Culture in the US* folder.

   A. Gullah
   B. Amish
   C. Cajun
   D. Puebloan peoples
   E. Marsh Arabs

One way folk culture still manifests itself in the modern world is through the diversity of food preferences of people around the world. These preferences have been shaped through a wide variety of environmental and cultural variables. Many religious groups, for example, have food taboos integrated into their religious texts. Open the *Food Taboos* folder and visit the five placemarked locations. Utilizing your text and any additional resources, determine likely food taboos for each location.

2. Identify the answer set that provides the most logical pairing of food taboos to the placemarked locations.

   A. 1 = Lobster, 2 = Beef, 3 = Fish, 4 = Horse, 5 = Pork
   B. 1 = Citrus, 2 = Horse, 3 = Pork, 4 = Milk, 5 = Lobster
   C. 1 = Beef, 2 = Fish, 3 = Horse, 4 = Dog, 5 = Pork
   D. 1 = Lobster, 2 = Pork, 3 = Wheat, 4 = Beef, 5 = Fish
   E. 1 = Pork, 2 = Cat, 3 = Dairy, 4 = Beef, 5 = Bird

Double-click the *McDonald's Global Headquarters* placemark to view the company's Illinois headquarters. While it is easy to associate fast food, and McDonald's in particular, with cultural homogenization, the chain does accommodate local preferences into their menus. Click the link associated with the *McDonald's Global Headquarters* placemark and explore some of the national websites for McDonald's. Pay attention to the wide variation in menu items offered at each location.

3. Based on your evaluation of the McDonald's Web sites identify the statement that is <u>not</u> supported.

    A. South Africa has a menu that is quite similar to a US McDonald's.
    B. The Chicken Maharaja Mac is the Big Mac equivalent in India.
    C. There is greater penetration of McDonald's into European markets than African markets.
    D. Across Saudi Arabia, McDonald's serves pure Halal meat.
    E. All of the McDonald's Web sites provide information in English.

Building styles represent one more way we can see culture manifested. The varieties of folk and popular architecture are remarkably diverse. Some regions have distinctive styles that permeate the built environment while others have an endless array of styles alongside one another. Open the *Folk Architecture* folder. Begin with *Santa Fe, NM*. Zoom in to Street View and navigate around the neighborhood. For Kinderdijk, use the same approach you did with Santa Fe to see the Street View, and then proceed to the other destinations. For the Abu Dhabi placemark, turn on the *Photos* layer and view a sample of the images from the site.

4. After viewing the sites in the *Folk Architecture* folder, identify the statement that is <u>not</u> supported.

    A. The mosque at Abu Dhabi appears to be a newly constructed building using folk architectural elements.
    B. The home in Aswan appears to have used locally available building materials.
    C. The Santa Fe neighborhood demonstrates that folk cultural-inspired landscapes can be seen in the contemporary United States.
    D. The Osaka skyscrapers are a better example of popular culture than folk culture.
    E. The buildings shown in Kuala Lumpur incorporate some folk cultural aspects (incorporation of Muslim art motifs) into their modern appearance.

Exploration 6.2 SHORT ESSAY

1. Explore the McDonald's national Web sites further. Identify three items from three different countries that reflect the incorporation of local food preferences into McDonald's menus.

    _____
    _____
    _____
    _____
    _____
    _____
    _____

2. Open the *Folk Agriculture* folder and view the five sites identified by red polygons. Which of the sites is not a good example of folk agriculture and why? Provide a very brief description of the type of agriculture associated with each of the other sites. Utilize outside resources to help explain the agricultural types if necessary.

_____

_____

_____

_____

_____

_____

_____

Exploration 6.3: Dispersed Popular

In recent decades, the internet has been one of the most important factors in accelerating the diffusion of popular culture. Open the *Dispersed Popular* folder and turn on the *Internet Users* folder. Now, study the global pattern of internet users. Divide the number of internet users in a country by its total population to calculate a measure of per capita internet usage.

Exploration 6.3: MULTIPLE CHOICE

1. Identify the statement that is supported by the *Internet Users* folder.

    A. China has the highest per capita internet usage in the world.
    B. The countries of Central Asia have generally higher numbers of total internet users than European countries.
    C. The United States has the highest per capita internet usage in the world.
    D. China has more internet users than the United States.
    E. Nigeria has one of the ten lowest internet user figures in the world.

Turn off the *Internet Users* folder. The World Heritage Committee works with countries around the world to identify and protect locations that are deemed to possess global natural or cultural attributes of global significance. The World Heritage Committee overwhelmingly recognizes locations of historic significance and/or folk cultural sites over those with a focus on popular culture. Five World Heritage sites have been placemarked in the *World Heritage* folder of which four represent contemporary or popular themes of culture. Turn on the *3D Buildings* layer and use clues such as the *Photos* layer to help you determine why each site has been recognized. You can also visit the World Heritage Web page, where you can find detailed descriptions of each site.

2. Identify the theme <u>not</u> associated with one of the placemarked World Heritage sites.

    A. The first nuclear bomb used in war.
    B. A railway built in very challenging mountainous conditions.
    C. An architectural style known as Bauhaus.
    D. The architecture of an opera hall.
    E. The artwork of Sigmar Polke.

Popular housing style is one facet of popular culture that can change through time. Many neighborhoods have a certain popular style of housing reflective of the era in which it was built.

Fly to the *Popular Housing* placemark. Zoom in to Street View and examine the style of popular housing that is prevalent in this neighborhood. If necessary, research housing styles on the internet.

3. The style of popular housing that is most common in the *Popular Housing* neighborhood is

    A. Ranch.
    B. Mansard.
    C. Neo-French.
    D. Contemporary.
    E. Split-Level.

Sports are a great example of the diffusion of popular culture around the world. Athletes like Serena Williams, Lionel Messi, and Lebron James are known globally as their sports are broadcast into people's living rooms from one corner of the world to another. Popular sports can also greatly impact local economies. Revenue is generated not only for the teams through ticket sales, but also for surrounding merchants of team merchandise on the Web. Turn on the *3D Buildings* layer and then open the *Popular Sports Stadia* folder and visit the placemarked stadia. Use clues such as the country location of each stadium to determine the popular sport played in each stadium. When you have completed your assessment, turn off the *3D Buildings* layer.

4. Match the popular sport associated most strongly with each stadium.

    A. 1 = Field hockey, 2 = Cricket, 3 = Australian Rules football, 4 = Football/Soccer, 5 = Lacrosse
    B. 1 = Football/Soccer, 2 = Rugby, 3 = Cricket, 4 = American football, 5 = Baseball
    C. 1 = Cricket, 2 = Rugby, 3 = American football, 4 = Baseball, 5 = Football/Soccer
    D. 1 = Baseball, 2 = American football, 3 = Cricket, 4 = Rugby, 5 = Football/Soccer
    E. 1 = Lacrosse, 2 = Football/Soccer, 3 = Rugby, 4 = American football, 5 = Baseball

Exploration 6.3 SHORT ESSAY

1. Uluru-Kata Tjuta National Park in Australia is one of the rare World Heritage sites that has been recognized for both its natural and cultural attributes. It includes huge red geological formations that form part of the belief system of the Anangu Aboriginal people of Australia. This is one of the best examples of the World Heritage Committee recognizing folk culture. Go to the *Uluru* placemark and explore the site. Turn on the *Photos* layer and examine some of the development in the vicinity. Describe your findings. Do you also see elements of popular culture? Do you think this is an encroachment upon the folk characteristics of the site?

    _____
    _____
    _____
    _____
    _____
    _____
    _____
    _____

2. Rugby is an extremely popular sport in New Zealand. The national team, known as the All-Blacks, is a perennial power in the realm of international rugby. Research the routine, called the Haka, that the All-Blacks perform prior to their matches. Explain how this is an example of folk and popular culture coming together.

_____

_____

_____

_____

_____

_____

_____

Exploration 6.4: Popular Globalization

The global penetration of popular culture has not been welcomed by some individuals, societies, and governments. The dilution of traditional values and cultures is viewed as a perilous threat in some parts of the world. Access to elements of popular culture can be limited by barriers put in place by governments. One way to get a general measure of potential barriers to popular culture penetration is to view Freedom House's assessment of Freedom in the World. Open the link to _Freedom House_ contained in the _Popular Globalization_ folder. Find the most recent year's Freedom in the World assessment in the "Publications" section of the Freedom House Web site. You will have better luck if you open this site in a browser outside of Google Earth™. When you click on an individual country, you will see a "Political Rights Score" and a "Civil Liberties Score." Higher scores indicate less freedom. For this assessment, note the civil liberties scores as they are a better reflection of governmental barriers to cultural diffusion.

Exploration 6.4: MULTIPLE CHOICE

1. Based on your evaluation of Freedom in the World civil liberties scores, identify the statement that is not supported.

    A. Myanmar (Burma) has very limited civil liberties.
    B. India has a better civil liberties rating than China.
    C. Venezuela has one of the lower civil liberty scores in the Americas.
    D. Sudan has very limited civil liberties.
    E. Sub-Saharan Africa has no countries classified as "Free" by Freedom House.

More specifically, the organization "Reporters without Borders" identifies "Internet Enemies" where access to information on the Web is severely curtailed. In many countries, internet access can be one of the principal avenues for diffusion of culture. Thus, a government limitation on the dissemination of information can slow the spread of cultural diffusion.

2. Click on the Reporters without Borders link. Now, scroll to the bottom of the Reporters without Borders Web page and use the information on countries identified as "Internet Enemies" to evaluate the following statements. Select the statement that is not supported by Reporters without Borders information.

    A. China and Google had a contentious relationship in 2010.
    B. Individual Web connections in Turkmenistan were only authorized in 2008.
    C. Cubans are guaranteed access to the internet as a socialist right, but some content is limited.
    D. In North Korea, the majority of the population does not even know that the internet exists.
    E. Searches with the word "Tiananmen" are sometimes blocked in China.

In the United States, there is little in the way of government-imposed barriers to cultural diffusion. In fact, some believe that popular culture has created monotonous cultural landscapes. Sometimes this is referred to as placelessness or the "geography of nowhere." Increasingly, one can visit newly developed areas across the United States and see the same familiar features, thus diminishing the influence of location. Open the *Geography of Nowhere* folder and tour some of the representative features of uniform landscapes.

3. Which of the following statements is supported by the *Geography of Nowhere* placemarks?

    A. American brands have not become prominent outside the United States.
    B. There is evidence of non-Western brands becoming prominent in the Unites States.
    C. American products have yet to become established in the Muslim world.
    D. Fast food restaurants are not a good example of placelessness.
    E. Big box stores cater to consumers traveling by foot rather than those traveling by automobile.

Use the historical imagery capability of Google Earth™ to study the locations of the *Geography of Nowhere* placemarks. Historical imagery is turned on by clicking the button with a clock icon that is found above the 3D Viewer.

4. Which of the following statements is not supported by historical imagery applied to the *Geography of Nowhere* placemarks?

    A. As late as 2000, the area of the *Big Box Stores* was undeveloped agricultural land.
    B. Through 2002, the hillside home of the *McMansions* placemark was undeveloped land.
    C. The home of the *Sanrio Hello Kitty Store* was an open field until 2002.
    D. The Taco Bell seen in the *Generic Building Forms* Street View was not built until at least 2010.
    E. The three miles along the main road (Hwy 3) north and west of the *Sprawling Commercial Development* placemark have seen significant additions in commercial development since 1990.

Exploration 6.4 SHORT ESSAY

1. Think about the resources needed for the production of the popular culture forms highlighted in the *Geography of Nowhere* folder. Note that most of the changes occurred in less than a ten year period. Comment on the potential impacts upon the globe's physical environments accompanied by this level of growth.

_____
_____
_____
_____
_____
_____
_____

2. The consideration of nature has not been entirely divorced from popular culture. Verify that the Historical Imagery is turned off and go to the *Yellowstone* placemark and study the scene. Think about the intersection of the natural world and popular culture, and provide a commentary.

_____
_____
_____
_____
_____
_____
_____

Encounter Human Geography

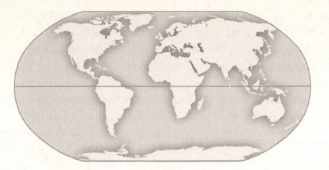

Name: _____

Date: _____

Chapter 7:
Language

Language is an essential element of cultural geography. It represents not only our primary means of communication, but also helps define our identities and express the relationship between groups of people and their natural environment. While the world contains thousands of living languages, many are related, like branches on a tree. Despite the staggering variety of languages, increasing homogeneity is the rule as English grows in prominence. But even within the growing influence of English, languages continue to change and evolve. The following explorations will provide you with an enhanced understanding of the world's tongues.

Download EncounterHG_ch07_Language.kmz from **www.mygeoscienceplace.com** *and open in Google Earth™.*

Exploration 7.1: English

The United States has the highest number of English speakers who speak the language as their first language. However, when the total number of English speakers in a country is measured (including those who speak the language as a first, second, or third language), a different picture emerges. Open the *Language* folders and view the five countries that have been placemarked in the *English Speakers* folder contained in the *English* folder.

Exploration 7.1 MULTIPLE CHOICE

1. Use your personal knowledge and logic combined with your text and any outside resources to identify the statement that is most likely to be true.

   A. Country 1 would likely be included in the top ten list of English as a first language but not included in the top ten for total English speakers.
   B. Country 2 would likely be included in both the top ten list of English as a first language and the top ten for total English speakers.
   C. Country 3 would likely be included in the top ten list of English as a first language but not included in the top ten for total English speakers.
   D. Country 4 would likely be included in both the top ten list of English as a first language and the top ten for total English speakers.
   E. Country 5 would likely be included in both the top ten list of English as a first language and the top ten for total English speakers.

Collapse and turn off the *English Speakers* folder. When one considers the expanse of English in relation to the historical and contemporary activities of the world's leading military and economic powers, it's not surprising to find English in every corner of the world. However, this is a dynamic situation. The predominance of English both waxes and wanes and undergoes unique geopolitical transformations. Take, for example, the country of India. Since the period of British imperialism, India continues to rename its cities. In some cases, a city will revert to its pre-British name. This may be an attempt to correct an English language spelling or to rename the city from its Anglo-oriented name to one of Indian origin. View the placemarked cities in the *Indian Names* folder by visiting the country of India in Google Earth™.

2. Using Google Earth™ and any outside research, determine the names of the placemarked cities and which of these place names does not reflect the legacy of British influence.

   A. Calcutta
   B. Mumbai
   C. Madras
   D. Bangalore
   E. Simla

Collapse and turn off the *Indian Names* folder. Now let's look at how language and environment influence one another. Oftentimes, the vocabulary for a given language develops partially as a response to the local environment. For example, a humid region, such as the southeastern United States, has numerous names in English for a flowing water body including creek, stream, run, and bayou. A rich vocabulary for place features is also more likely to arrive in regions with a high degree of language diversity compared to an area where only one language is spoken. Examine the landscapes of the five placemarks in the *Vocabulary and Habitat* folder.

3. Which of the following landscape features would be most likely to have multiple names for the setting in which it is associated?

   A. Glacier
   B. Desert
   C. Stream
   D. Hill
   E. Plain

Collapse and turn off the *Vocabulary and Habitat* folder. While English may seem to be taking over the world, there is still a tremendous degree of language diversity found around the globe. This diversity can be witnessed not only by language differences, but by the many dialects found within languages. For example, linguists have identified up to 24 different dialects of American English. The Public Broadcasting Service has created a program and interactive Web site called "Do You Speak American?" Open the link associated with the *Unique American Dialect* placemark to view the Web sites information on some of the American dialects. Read about the different American dialects on the Web site.

4. Which of the following dialects is most strongly associated with the region immediately surrounding the Unique American Dialect placemark?

   A. *A*-Prefixing
   B. Cajun
   C. Chicano
   D. Lumbee
   E. Pittsburghese

Exploration 7.1 SHORT ESSAY

1. Open the *Dialect Differences* folder and go to the three marked locations. Now think about how you envision the dialects/accents of people from each of the regions. Think of a popular reference for each location (such as a television show, film, or singer) that you think has the characteristic dialect of the placemarked locations. Provide and explain your examples.

   _____
   _____
   _____
   _____
   _____
   _____
   _____

2. Toponyms are names given to places on Earth. For example, the name of your hometown is a toponym. Toponyms can give all kinds of interesting cultural clues from who settled an area to the environmental perception of a location to significant historical events. Think about some toponyms that you are familiar with and identify and briefly discuss three toponyms from places you have lived or currently live.

   _____
   _____
   _____
   _____
   _____
   _____

Exploration 7.2: Languages Related to English

Collapse and turn off the *English* folder. English is but one of many languages contained in the Indo-European language family—a collection of languages that are related through a common ancestor. Indo-European languages are the most extensively spoken on the planet. Visit the locations placemarked in the *Indo-European* folder contained in the *Languages Related to English* folder.

Exploration 7.2 MULTIPLE CHOICE

1. Google™ to learn more about where one finds speakers of Indo-European languages. Identify the location where you would be least likely to engage in a conversation in an Indo-European language if speaking with a native person.

   A. A
   B. B
   C. C
   D. D
   E. E

Collapse and turn off the *Indo-European* folder. Ethnologue is a language resource available on the Web that catalogs the more than 6,000 living languages of the world. Use the *Ethnologue* link in Google Earth™ to open the Web site. Then, click on the country Index with Maps link and explore data regarding the languages of Cameroon.

2. Identify the statement that is supported by the data available from Ethnologue.

   A. More than 400 distinct languages are spoken in Cameroon.
   B. Fulfulde is spoken mainly in the south of the country.
   C. English is one of the official languages of Cameroon.
   D. Ngiemboon has been identified as a language in danger of extinction in Cameroon.
   E. French is not spoken in Cameroon.

Romance languages form one of the branches of the Indo-European family. These languages have diffused across the globe via the mechanisms of colonialism, trade, and communication. Look at the countries placemarked in the *Colonial Power* folder. These countries all share a common colonial past and subsequently a common language thread to this day.

3. The countries contained in the *Colonial Power* folder had a colonial association with

   A. Belgium.
   B. England.
   C. France.
   D. Germany.
   E. Portugal.

Some Romance languages have more speakers in countries other than their namesake. For example, the most number of Portuguese speakers in a single country is in Brazil rather than Portugal. In the following analysis, consider speakers to include fluent and partial speakers of each language.

4. Each placemarked country in the *Romance* folder is one of the top five countries for a respective Romance language. Match the placemarked numbers in the *Romance* folder with the correct list below.

   A. 1 = Spanish; 2 = French; 3 = German; 4 = Italian; 5 = Portuguese
   B. 1 = German; 2 = Italian; 3 = Spanish; 4 = Portuguese; 5 = French
   C. 1 = Portuguese; 2 = French; 3 = German; 4 = Italian; 5 = Spanish
   D. 1 = Italian; 2 = Spanish; 3 = Spanish; 4 = German; 5 = Portuguese
   E. 1 = French; 2 = Spanish; 3 = Portuguese; 4 = German; 5 = Italian

Exploration 7.2 SHORT ESSAY

1. Language has been a point of contention in any number of geopolitical contexts throughout history. Fly to the *Language Divides* placemark for a modern example. Utilize outside research to understand the language division at this location and provide a brief summary. Then, turn off the *Language Divides* placemark.

    _____
    _____
    _____
    _____
    _____
    _____
    _____

2. Another contemporary example that nearly led to the split of a country is found at the *Language Divides 2* placemark. What was at issue at this placemark? Identify at least two changes to the structure of the country in question as a result of the tensions associated with the language divide.

    _____
    _____
    _____
    _____
    _____
    _____
    _____

Exploration 7.3: Other Language Families

The second largest language family behind Indo-European is the Sino-Tibetan family. This language family uses ideograms, which represent concepts or ideas, rather than specific sounds. Google Earth™ shows the ideograms for the Chinese cities alongside the English toponyms.

Exploration 7.3: MULTIPLE CHOICE

1. Searching Google Earth™ in English, identify the choice below that displays the ideograms for the city of Chengdu.

    A. 成都市
    B. 赣州市
    C. 深圳市
    D. 舟山市
    E. 丹东市

After the Indo-European and Sino-Tibetan families, the next largest language families include Niger-Congo, Afro-Asiatic, Austronesian, Dravidian, and Altaic.

2. Which of the following options provided below has correctly identified each country with its proper language family?

    A. Gabon = Niger-Congo, Yemen = Austronesian, Indonesia = Afro-Asiatic, southern India = Dravidian, Turkey = Altaic
    B. Gabon = Altaic, Yemen = Austronesian, Indonesia = Afro-Asiatic, southern India = Dravidian, Turkey = Niger-Congo
    C. Gabon = Austronesian, Yemen = Altaic, Indonesia = Dravidian, southern India = Afro-Asiatic, Turkey = Niger-Congo
    D. Gabon = Niger-Congo, Yemen = Afro-Asiatic, Indonesia = Austronesian, southern India = Dravidian, Turkey = Altaic
    E. Gabon = Dravidian, Yemen = Altaic, Indonesia = Niger-Congo, southern India = Afro-Asiatic, Turkey = Austronesian

3. Collapse and close the *Families* folder. Go to the Google Translate™ page. There is a link in the *Translations* folder. Use the translator tool to identify the selection that is not appropriate for a descriptive statement about the scene in the native language. Be sure to listen to each translation in its native language as well. Hints: Copy and paste the sentence from the online component of the workbook into Google Translate™. Be sure you have selected "Detect language." Each sentence begins "This is a…"

    A. Это река.
    B. Ini adalah seorang koboi.
    C. Tämä on saari.
    D. Es tracta d'un llac.
    E. 这是一座山.

Collapse and turn off the *Translations* folder. There are numerous places in the United States where one can see non-Indo-European languages on display. In addition to Native American communities, ethnic neighborhoods in larger cities can be some of the best places to witness this. Open the *In the US* folder and explore the Street View scenes. Be sure to navigate a surrounding block or two at the street level.

4. Based on the Street View scenes found *In the US* folder, identify the statement that is supported.

    A. Chinese ideograms are not represented at any of the selected sites.
    B. Dorchester is associated with a Vietnamese community.
    C. Chinatown is highlighted in the Los Angeles example.
    D. Each of the sites shows a total separation between English language signs and non-English language signs.
    E. Japantown appears to have undergone the lowest degree of gentrification.

Exploration 7.3 SHORT ESSAY

1. Can you think of a non-spoken language? If so, what population uses it?

_____
_____
_____
_____
_____
_____
_____

2. In one question above you had the opportunity to listen to translations of short sentences describing geographic features. One of the sentences was spoken using an Indo-European language. Which one do you think it was and why?

_____
_____
_____
_____
_____
_____
_____

Exploration 7.4: Endangered and Emerging Languages

Turn off and collapse the *Other Language Families* folder. Open the *Endangered and Emerging Languages* folder. Language diversity is generally decreasing around the world. There are numerous endangered languages because they are only spoken by a limited number of older speakers and are not actively being taught. Recent centuries have demonstrated that languages can come back from near extinction. One example is marked with the *Surviving Language* placemark at the location at which it is most strongly associated. Complete outside research if necessary to assist with this question.

Exploration 7.4 MULTIPLE CHOICE

1. Select the most accurate statement regarding the *Surviving Language* placemark.

   A. Judaism is a language that was invented here in 1949.
   B. The ongoing conflicts between Israelis and Palestinians led to the reestablishment of Arabic here in the twentieth century.
   C. People that came from all around the world to form Israel helped reestablish Hebrew.
   D. Israel established Aramaic as its official language when it was created.
   E. Israel has no official language because of its population of both Jewish and Arab citizens.

An isolated language is one that is unrelated to any other language family. Look at the examples included in the *Isolated Language* folder.

2. Using outside resources, identify the isolated language within the *Isolated Language* folder that best represents an isolated language.

   A. Catalan
   B. Etruscan
   C. Basque
   D. Welsh
   E. Romansh

Collapse and turn off the *Isolated Language* folder. While English is ever more present, it continues to transform and develop unique dialects along the way. As new spellings, vocabulary, and pronunciations are added based on local tendencies, the language can develop unique local flavor. English, for example, has increasingly become hybridized as it blends and fuses with other languages. The creation of Spanglish, Denglish, Ebonics, Franglais, and Singlish are all examples of the continual transformation of the English language. Google™ each of these dialects to find out how and where they came into being.

3. Assign the dialects to the most appropriate placemarked location from the *Evolving English* folder.

   A. 1 = Singlish, 2 = Denglish, 3 = Franglais, 4 = Spanglish
   B. 1 = Denglish, 2 = Singlish, 3 = Spanglish, 4 = Franglais
   C. 1 = Singlish, 2 = Spanglish, 3 = Denglish, 4 = Franglais
   D. 1 = Denglish, 2 = Singlish, 3 = Franglais, 4 = Spanglish
   E. 1 = Franglais, 2 = Singlish, 3 = Spanglish, 4 = Denglish

Turn off and collapse the *Evolving English* folder. The Living Tongues Institute has identified a number of global language "hotspots" in order to better understand the distribution of global linguistic diversity, and the threat of extinction, as well as to prioritize issues for further research. Explore the Web site associated with the *Language Hotspots* folder.

4. Identify the region in the *Language Hotspots* folder that is least representative of the language hotspots identified on the linked National Geographic map.

   A. Oklahoma/Southwest
   B. Northern Australia
   C. Eastern Siberia
   D. Northern Africa
   E. Central South America

Exploration 7.4 SHORT ESSAY

1. "English-Only" laws have become increasingly commonplace in the United States in the last two decades. Proponents of these laws suggest that it is too expensive to provide official documents in multiple languages and English-only laws encourage immigrants to become assimilated in the host society. On the other hand, opponents believe the laws are racist and serve to deny basic services and information to people who do not understand English. Three states that have not passed English-only laws are identified in the *No English-only Law* folder. Using Google™, provide some insight as to potential reasons why English-only laws have not been enacted in these states.

    _____
    _____
    _____
    _____
    _____
    _____
    _____

2. Certain languages are associated with certain religions. For example, Islam is strongly tied to Arabic and Hebrew is tied to Judaism. As you think about the future spread of religion around the world, what do you think some impacts could be in the realm of language? For example, do you think there could be a connection between the most rapidly growing religious traditions and languages associated with them?

    _____
    _____
    _____
    _____
    _____
    _____

Encounter Human Geography

Name: _____

Date: _____

Chapter 8:
Religion

Religion represents a part of cultural identity that many people of the world value over all else. Religion contributes to some of the starkest cultural differences. Value systems of individuals and groups are shaped by religious traditions. Many of the world's most striking structures are religious in nature. Let's take a look at the origin, distribution, and diffusion of global religions; visit a few selected holy places and spaces; and investigate the violent side of religion.

Download EncounterHG_ch08_Religion.kmz from **www.mygeoscienceplace.com** *and open in Google Earth™.*

Exploration 8.1: Distribution of Religions

Expand both the *Religion* folder and the *Distribution of Religions* folder. Each of the labeled countries in the *Highest Percentage Countries* folder contains the most concentrated religious populations in the world according to Adherents.com. For example, the population of Iceland is cited as 94 percent Lutheran. It has the highest Lutheran percentage in the world. Peruse the labeled countries in the *Highest Percentage Countries* folder and Google™ the religions and/or countries included in this question to gain an understanding of the relevant religions' distributions of adherents.

Exploration 8.1 MULTIPLE CHOICE

1. Turn on the *Borders and Labels* folder and match the labeled countries with the religions (or lack of religion) that reach their highest level of saturation in each respective country.

   A. 1 = Atheism/Agnosticism, 2 = Sikhism, 3 = Baha'i, 4 = Hinduism, 5 = Buddhism
   B. 1 = Lutheranism, 2 = Sikhism, 3 = Islam, 4 = Hinduism, 5 = Buddhism
   C. 1 = Theravada Buddhism, 2 = Evangelical Christian, 3 = Rastafarianism, 4 = Islam, 5 = Baha'i
   D. 1 = Catholicism, 2 = Evangelical Christian, 3 = Islam, 4 = Hinduism, 5 = Baha'i
   E. 1 = Lutheranism, 2 = Unitarianism, 3 = Islam, 4 = Hinduism, 5 = Tenrikyo

Reading the cultural landscape can often provide conspicuous clues as to the prevailing religious tendencies of a location. Turn off the *Highest Percentage Countries* folder and turn on the *3D Buildings* layer. Then, double-click on each structure listed in the *Built Reflections of Culture* folder, and visit these representative places of worship for a variety of religious traditions. Google™ the structures included in the following question.

2. Identify the structure that is correctly labeled.

    A. Muslim mosque
    B. Catholic cathedral
    C. Buddhist temple
    D. Kinto shrine
    E. Eastern Orthodox church

Have you ever heard of a "diaspora"? The term refers to a collective group of people that has been dispersed outside its original homeland. Different religions have been spread out across the globe, creating distinctive distributions. Turn off the *3D Buildings* layer and open the *Diaspora* folder. Utilize your text and any outside resources to research the locations of the following groups' adherents.

3. What religious group's diaspora is most likely represented by the placemarked cities in the *Diaspora* folder?

    A. Jain
    B. Jewish
    C. Shi'a Islam
    D. Protestantism
    E. Lamaist Buddhism

Islam is the fastest growing religion in the world. An essential element (one of the "Five Pillars of Islam") of the religious tradition is centered on the location associated with the *Islam Pillar* placemark. Turn on the *3D Buildings* layer and double-click the *Islam Pillar* placemark. The enormous skyscraper complex is, in fact, a Muslim-only hotel that has recently been constructed to accommodate visitors to the site. Read about the Five Pillars of Islam in your textbook or on the Web.

4. What Pillar of Islam is most closely tied to this location?

    A. Shahadah
    B. Salat
    C. Sawm
    D. Zakat
    E. Hajj

Exploration 8.1 SHORT ESSAY

1. Africa is undergoing some of the most radical changes in terms of wholesale religious transitions/conversions. Research this phenomenon in your text and outside resources and then examine Google Earth™ for visual evidence. For example, you could locate a city along the boundary of Christianity and Islam with both large mosques and churches. Provide the location of your example (lat/long) in your response.

_____

_____

_____

_____

_____

_____

2. Examples of the presence of a given religion within the cultural landscape come in a widely varying array of places of worship. Regional differences can reflect more than variation in prevailing religious preference. The structures can also illustrate folk and/or popular culture by their architectural styles and choice of building materials. Use Google Earth™ to locate one example of a place of worship with heavy folk cultural influence and another with predominant popular culture influence.

_____

_____

_____

_____

_____

_____

_____

Exploration 8.2: Origin and Diffusion of Religion

Most of the world's religions and branches and denominations of religions have strong associations with "place." The birthplaces of founders, locations of administrative headquarters, or sites of miracles, are a few examples. Turn off the *Diaspora* folder and open the *Origin and Diffusion of Religion* folder. Then, watch the *Religious Migration* tour to see a number of significant places for one branch of a major monotheistic tradition.

Exploration 8.2 MULTIPLE CHOICE

1. Identify the answer choice that is most strongly associated with the *Religious Migration* tour.

A. Saint Mark
B. Dalai Lama
C. Baha'u'llah
D. Ali
E. Joseph Smith, Jr.

Different religions have spread far and wide across the globe. Sometimes one may encounter pockets of believers in locations that one might otherwise not expect to be there. Five locations of significant Muslim populations are placemarked in the *Why Islam* folder. Open this folder and research these locations to gain an understanding of the factors that have contributed to higher Muslim populations at these locations.

2. Select the best association of terms with locations placemarked in the *Why Islam* folder.

    A. 1 = Crusades, 2 = Francophone, 3 = Holocaust, 4 = Malcolm X, 5 = Zanzibar
    B. 1 = Treaty of Tordesillas, 2 = Arab traders, 3 = Holocaust, 4 = Ottoman, 5 = Zion
    C. 1 = Colonialism, 2 = Gastarbeiter, 3 = Zanzibar, 4 = Wallace Muhammad, 5 = Francophone
    D. 1 = Arab traders, 2 = Colonialism, 3 = Gastarbeiter, 4 = Ottoman, 5 = Wallace Muhammad
    E. 1 = Colonialism, 2 = Arab traders, 3 = Elijah Muhammad, 4 = Shiite, 5 = Francophone

The United States has strong regional variation when it comes to religion, as certain religious traditions are more strongly associated with particular regions, cities, or parts of cities than others. Visit the locations placemarked in the *Global Connections* folder.

3. Determine the best match between the US places of worship and the selected religious focal points.

    A. Boston = Jerusalem, Fargo = Holy See, Miami = Jerusalem, New Orleans = Scandinavia, Seattle = Vietnam
    B. Boston = Holy See, Fargo = Scandinavia, Miami = Jerusalem, New Orleans = Haiti, Seattle = Vietnam
    C. Boston = Scandinavia, Fargo = Jerusalem, Miami = Vietnam, New Orleans = Holy See, Seattle = Haiti
    D. Boston = Haiti, Fargo = Jerusalem, Miami = Vietnam, New Orleans = Holy See, Seattle = Scandinavia
    E. Boston = Vietnam, Fargo = Haiti, Miami = Jerusalem, New Orleans = Scandinavia, Seattle = Holy See

Some religions can take unexpected paths to diffusion. For example, one branch of religion remained less well known until its leader fled into exile in the wake of a dramatic political and social upheaval in the mid-twentieth century. Subsequently, this leader has spoken to millions around the world about the situation in his homeland, which has greatly contributed to awareness of this religious tradition. The placemarked locations in the *Religion in Exile* folder represent important places in this tradition. Turn on the *3D Buildings* layer to better view the second site.

4. What religious tradition (discussed above) currently operates with its administrative headquarters in exile?

    A. Orthodox Christianity
    B. Hinduism
    C. Sunni Islam
    D. Tantrayana Buddhism
    E. Sikhism

Exploration 8.2 SHORT ESSAY

1. Go to the *Venice Ghetto* placemark. Explore the site using the *3D Buildings* layer and the *Photos* layer. Be sure to examine the 360° view in the *Photos* layer. This site is one of Europe's earliest Jewish ghettos. If you are unfamiliar with the term *ghetto*, research it on the Web. Based on your visual appraisal of the site, what are some clues that this could have been a ghetto?

_____
_____
_____
_____
_____
_____

2. The calendar has played a large role in many religions, as it is tied with the changing of seasons and significant environmental transitions related to agriculture. Early religious authorities utilized knowledge of the celestial cycles to cement their power amongst their followers. They would not have had the monopoly on knowledge if tools like the internet and Google Earth™ had existed. Google Earth™ provides a wide array of opportunities for celestial analysis by clicking the button that enables you to switch between the Earth, sky, Mars, and the moon. The button looks like a small planet with a ring around it. Explore the information and imagery associated with the Earth, sky, Mars, and the moon, and provide a comment on how this could have been useful to a person in the historic past.

_____
_____
_____
_____
_____
_____

Exploration 8.3: Holy Places and Spaces

Religions' intimate connection with place and space is best seen in the vast collection of places and spaces deemed holy by the world's religious traditions. The birth places of the founders of some of the world's religions are excellent examples of some of the holiest of sites. Open the *Holy Places and Spaces* folder and then the *Birth Places* folder and visit the placemarks. Research the birth locations of the founders of the religions included in the following question.

Exploration 8.3 MULTIPLE CHOICE

1. Identify the religion associated with each birth place in the *Birth Places* folder.

   A. 1 = Islam, 2 = Christianity, 3 = Buddhism, 4 = Sikhism, 5 = Baha'i Faith
   B. 1 = Judaism, 2 = Sikhism, 3 = Baha'i Faith, 4 = Hinduism, 5 = Christianity
   C. 1 = Buddhism, 2 = Baha'i Faith, 3 = Sikhism, 4 = Christianity, 5 = Hinduism
   D. 1 = Christianity, 2 = Hinduism, 3 = Christianity, 4 = Baha'i Faith, 5 = Islam
   E. 1 = Sikhism, 2 = Islam, 3 = Christianity, 4 = Hinduism, 5 = Baha'i Faith

A number of locations around the world have gained prominence in their religious traditions long after the establishment of the religion. Perhaps there was a major philosophical shift, schism in the religion, or an alleged miracle. Turn on the *3D Buildings* layer. In the *Evolving Religion* folder, there are five places marked that correspond with descriptive clues. For the placemarked locations, you may need to consider the wider geographic context, such as the city or country in which the placemark is located. Utilize the Web to search for additional information if necessary.

2. Match the clues to their appropriate placemarked location from the *Evolving Religion* folder.

   A. 1 = split over slavery, 2 = Abrahamic significance, 3 = 95 Theses, 4 = religious revival in the last 20 years after a change in political ideology, 5 = cathedral → mosque → museum

   B. 1 = Abrahamic significance, 2 = split over slavery, 3 = religious revival in the last 20 years after a change in political ideology, 4 = cathedral → mosque → museum, 5 = 95 Theses

   C. 1 = religious revival in the last 20 years after a change in political ideology, 2 = 95 Theses, 3 = cathedral → mosque → museum, 4 = split over slavery, 5 = Abrahamic significance

   D. 1 = cathedral → mosque → museum, 2 = 95 Theses, 3 = Abrahamic significance, 4 = religious revival in the last 20 years after a change in political ideology, 5 = split over slavery

   E. 1 = 95 Theses, 2 = cathedral → mosque → museum, 3 = split over slavery, 4 = Abrahamic significance, 5 = religious revival in the last 20 years after a change in political ideology

3. In today's world, many religious adherents travel to places of importance for their religious tradition. Pilgrimages to these holy places are not only important to the followers of religions, but also to the economies of the places that are visited. There are a number of religious pilgrimage sites in Europe. One of the pilgrimages is represented by the *Pilgrimage Symbol* contained in the *Pilgrimage* folder. View the locations placemarked in the *Pilgrimage* folder and determine the one you would most likely visit if you were on the pilgrimage associated with the *Pilgrimage Symbol*. You may want to utilize the capability of Google™ to search for images.

   A. A
   B. B
   C. C
   D. D
   E. E

Sacred places often have distinctive architectural elements associated with them. Familiarize yourself with the following terms: campanile, minaret, onion dome, spire, stupa. Open the *Religious Architecture* folder, turn on the *3D Buildings* layer, and tour the placemarked locations.

4. Match the architectural features with the locations placemarked in the *Religious Architecture* folder.

    A. 1 = campanile, 2 = spire, 3 = minaret, 4 = stupa, 5 = onion dome
    B. 1 = spire, 2 = campanile, 3 = minaret, 4 = onion dome, 5 = stupa
    C. 1 = stupa, 2 = minaret, 3 = onion dome, 4 = spire, 5 = campanile
    D. 1 = minaret, 2 = onion dome, 3 = stupa, 4 = campanile, 5 = spire
    E. 1 = onion dome, 2 = stupa, 3 = campanile, 4 = minaret, 5 = spire

Exploration 8.3 SHORT ESSAY

1. In Latin America, there are a number of roadside shrines and sanctuaries. Go to the *Santuario del Camionero* placemark and view the area surrounding this shrine. Use your assessment of the local geography along with Google Translator™ (*translate.google.com*) to determine why this shrine is located here.

    _____
    _____
    _____
    _____
    _____
    _____
    _____
    _____

2. Each of the sites (excluding the correct answer) highlighted in the *Pilgrimage* folder are destinations for religious pilgrims. Research each site and identify which one you think would be the most interesting to visit. Explain your answer.

    _____
    _____
    _____
    _____
    _____
    _____
    _____
    _____

Exploration 8.4: Religious Conflict

Conflict has proven to be an inherent part of many religious traditions through the ages. Sometimes this conflict is between religious groups while other times it is within religious traditions. Perhaps the ultimate focal point of religious conflict is the contested space of Jerusalem. Jerusalem is unique in that it contains a number of sacred sites for the world's big three monotheistic religious traditions: Christianity, Judaism, and Islam. Open the *Religious Conflict* folder and go to the *Jerusalem* placemark. Use Google Earth™ and any outside resources to identify the locations of the following: the location of Christ's crucifixion, the location of the western wall of the Second Jewish Temple, and the location of Muhammad's ascension.

Exploration 8.4 MULTIPLE CHOICE

1. Use the measure tool to determine how large a radius from the *Jerusalem* placemark is needed to include the following sites: the location of Christ's crucifixion, the location of the western wall of the Second Jewish Temple, and the location of Muhammad's ascension.

   A. 0.1 kilometer
   B. 0.5 kilometer
   C. 1 kilometer
   D. 3 kilometers
   E. 9 kilometers

2. Now visit the placemarks contained in the *Common Theme* folder. Utilize the *Photos* layer and any outside resources to learn about these sites. As a reminder, it's always a good idea to turn off the *Photos* layer when you are done using it because it clutters the view and will slow down the performance of Google Earth™.

   A. These are sites where the Apostle Paul assembled his armies to conquer Rome.
   B. These are Jewish castles from the time of Abraham.
   C. These are sites of Jewish-Christian battles from the sixteenth century.
   D. All three of these sites are associated with Hezbollah.
   E. These are fortified positions that were involved in the Crusades.

Acute religious violence has a long history in human affairs that continues to the present day. The placemarks in the *Violent Events* folder includes a selection of places associated with violence linked to religion. Utilize the *Photos* layer and any outside resources to learn about the events that occurred at these locations.

3. Match the sites with the most appropriate key words.

   A. 1 = Arkansas emigrants, 2 = Vernon Howell, 3 = Freedom of speech, 4 = George Tiller, 5 = People's Temple
   B. 1 = George Tiller, 2 = Vernon Howell, 3 = Freedom of speech, 4 = Arkansas emigrants, 5 = People's Temple
   C. 1 = Freedom of speech, 2 = George Tiller, 3 = Arkansas emigrants, 4 = People's Temple, 5 = Vernon Howell
   D. 1 = People's Temple, 2 = Freedom of speech, 3 = Vernon Howell, 4 = Arkansas emigrants, 5 = George Tiller
   E. 1 = Vernon Howell, 2 = Arkansas emigrants, 3 = George Tiller, 4 = People's Temple, 5 = Freedom of speech

The *Scotland 1* and *Scotland 2* placemarks are soccer stadiums in the city of Glasgow, Scotland. European soccer has a long history of violence, as well. Additionally, some teams have strong religious associations.

4. Perform any necessary outside research and identify the statement that is most strongly supported.

    A. Scotland 1 has a strong Protestant association and Scotland 2 has a strong Catholic association.

    B. These are stadiums for the Scottish Premier League, which only allows Christian athletes.

    C. St. Mirren supporters are not allowed in either of these stadiums because of their Catholic background.

    D. Scotland 1 is associated with Rangers and Scotland 2 is associated with Muslim supported Hibernian.

    E. These stadia are no longer in use because of religion-fueled soccer violence.

Exploration 8.4 SHORT ESSAY

1. Samuel P. Huntington coined the phrase "the bloody borders of Islam" to describe the disproportionate level of religious conflict found in areas that abutted Islam. One such example is Nigeria, where thousands have died in Christian-Muslim clashes in the last decade. *Maiduguri* has been a recent trouble spot with both churches and mosques burned to the ground. Do you believe there is substantive evidence to claim the borders of Islam are bloody? Explain your answer. Be sure to provide place examples in your response if appropriate.

    _____
    _____
    _____
    _____
    _____
    _____
    _____

2. The *Geopolitical Upheaval* polygon highlights an area that fractured in the 1990s into numerous states. While there were many contributing devolutionary forces in Yugoslavia, religion was certainly part of the equation. Provide another contemporary example where religious forces have contributed to the break-up or impending break-up of a country.

    _____
    _____
    _____
    _____
    _____
    _____
    _____

Name: _____

Date: _____

Chapter 9: Ethnicity

Ethnic groups, or people with common ancestry and cultural traditions, are the focus of this chapter. The distribution of ethnic groups around the world, in the United States, and in individual communities creates uniquely distinctive landscapes, cause for political recognition, and, sadly, the potential for conflict. We will investigate these elements in the following explorations of ethnicity.

Download EncounterHG_ch09_Ethnicity.kmz from **www.mygeoscienceplace.com** *and open in Google Earth™.*

Exploration 9.1: US Ethnic Distribution

Open the *US Ethnic Distribution* folder and individually view the *Ethnicity 1*, *2*, *3*, and *4* folders. Each of these folders displays large-scale ethnic patterns at the level of the state. Two of the ethnicities have been normalized by population. This means that the total number of persons in an ethnic group for each state was divided by the total population of each state to account for the wide range in state populations. A percentage measure of ethnicity is the result of normalization.

The other two ethnicities simply symbolize the gross number of persons in each group per state. Without normalization, it is likely that states with very large populations such as Texas and California will have higher values for ethnic groups than states with smaller populations such as Montana or Wyoming. Texas, for example, will have a much higher number of white persons than Montana. However, the percentage of white persons (normalized data) in Texas will be lower than that of Montana. For both the normalized and non-normalized groups, darker colors equate to higher values.

Exploration 9.1 MULTIPLE CHOICE

1. After viewing the four *Ethnicity* folders, determine what major ethnic group of the United States is not represented.

   A. African American
   B. Asian
   C. Hispanic
   D. Native American
   E. White

Following your evaluation of the ethnic folders, you should know what ethnic group is represented by each folder. Think about the relative size of each state's population compared to the population of ethnic groups.

2. Which of the large-scale ethnic patterns had not been normalized?

    A. Asian and African American
    B. Hispanic and White
    C. Hispanic and Native American
    D. Asian and Native American
    E. African American and White

Whenever we are looking at maps such as the state-level ethnicity folders, we must be sure not to make the assumption that this represents the smaller units of the state uniformly. For example, a state may have a very limited Hispanic population, but those persons may be clustered in one or two counties. These counties may therefore have normalized populations (percentage of the population in that county) much higher than state or even national averages. Turn off the *Ethnicity* folders. Turn on the *County Ethnicity* folder to see a county-level distribution of an ethnic group. Determine the group represented and then proceed with the question. You can use the Search pane to locate the counties if necessary.

3. Which of the following counties does not follow its state trend in regard to the ethnicity represented in the *County Ethnicity* folder?

    A. Tama County, Iowa
    B. Cherokee County, Oklahoma
    C. Santa Clara County, California
    D. Sunflower County, Mississippi
    E. Shannon County, South Dakota

Turn off the *County Ethnicity* folder. Ethnic groups can be mapped by variables other than a physical count of people. One such proxy measure is health, as certain groups are more prone to certain ailments. Turn on the *Stomach Cancer* folder to see a state-level map of locations with higher incidences of this disease.

4. Based on the *Stomach Cancer* map and any outside research, select the most accurate statement.
    A. Stomach cancer rates are high in states like North Dakota and New Mexico because of the high percentage of Hispanic persons.
    B. Stomach cancer is highest in the southeastern United States because of unhealthy diets and obesity.
    C. There are no regional patterns associated with stomach cancer.
    D. California has the lowest rate of stomach cancer because of its healthier lifestyle.
    E. Stomach cancer rates are higher in the upper Midwest as a result of Scandinavian food preferences.

Exploration 9.1 SHORT ESSAY

1. Provide an explanation that elaborates on your response to the question concerning stomach cancer.

_____
_____
_____
_____
_____
_____

2. Think of something else that could be used as a proxy measure for an ethnic group's presence.

_____
_____
_____
_____
_____
_____

Exploration 9.2: Ethnicity in the City

Ethnic diversity can manifest itself in the city in a number of ways, including ethnic restaurants, signs printed with non-native languages, and ethnically oriented shops and businesses. Open the *Ethnicity in the City* folder and then open the *Ethnic Neighborhood* folder. Examine the street views for clues as to the dominant ethnicity at each location.

Exploration 9.2 MULTIPLE CHOICE

1. Match the *Ethnic Neighborhood* placemarks with the most appropriate ethnicity.

   A. 1= Polish, 2 = Swiss, 3 = Vietnamese, 4 = Laotian, 5 = Indian
   B. 1 = Greek, 2 = Chinese, 3 = Thai, 4 = Italian, 5 = African American
   C. 1 = Greek, 2 = Italian, 3 = Chinese, 4 = Thai, 5 = Indian
   D. 1 = Italian, 2 = Swedish, 3 = Vietnamese, 4 = Greek, 5 = Thai
   E. 1 = Indian, 2 = Cuban, 3 = Italian, 4 = Thai, 5 = Greek

Turn off the *US Ethnic Distribution* folder and turn on the *Ethnic Cities* folder. The ethnic distributions of Baltimore, Miami, New Orleans, Philadelphia, and Spokane are represented by dot maps (based on the 2000 census) in the *Ethnic Cities* folder. Eric Fischer created these maps where red is White, blue is Black, green is Asian, orange is Hispanic, and gray is Other, and has generously made them available on Flickr under a Creative Commons license. Each dot represents 25 people and the data is from the 2000 US census. Examine the maps. Remember that you can adjust the transparency with the slider located at the bottom of the *Places* pane.

2. After studying the dot maps in the *Ethnic Cities* folder, identify which of the following statements is most accurate.

   A. In *New Orleans*, only whites live in areas immediately adjacent to the Gulf of Mexico.
   B. *Spokane* has the highest degree of ethnic diversity of the selected cities.
   C. The suburbs of *Philadelphia* have higher concentrations of African American residents than the city core.
   D. The core of *Baltimore* is disproportionately composed of African American persons, while the suburbs are predominantly white.
   E. The prime beachfront real estate of *Miami* is occupied primarily by Hispanic residents.

When the maps in the *Ethnic Cities* folder were brought into Google Earth™ from Flickr, they had to be adjusted in size and orientation to best approximate the location of ethnic groups in each city. Turn on and go to the *St. Louis* overlay. Select *St. Louis* in the table of contents, and then go to Edit, Properties at the top of the page. The Edit Image Overlay box will open and the gray square overlay will have green brackets. Use the brackets to adjust the overlay (stretch it, shrink it, rotate it with the Location tab) until it matches the underlying imagery in Google Earth ™. Remember you can adjust the transparency of the overlay using the slider, as well. Hint: The overlay will need to be resized <u>and</u> rotated. Use the river path as a guide.

Note: In the Mac version of Google Earth™ you will need to edit the overlay by clicking Edit and then clicking Get Info.

3. After adjusting the map to its appropriate location and size, identify which of the following statements is most accurate.

   A. The part of St. Louis east of the river has the highest concentration of African American residents.
   B. Southwest St. Louis is predominantly white.
   C. The highest density of Asian residents is found in the northern suburbs.
   D. The largest ethnic group other than white in St. Louis is Hispanic.
   E. There is a high correlation between the white population and riverfront property.

Now turn on and fly to the *Philadelphia Redlining* map. Redlining refers to the practice of denying certain services to certain geographic areas on the basis of race or ethnicity. While the term applies to any number of situations, it is often discussed in the context of areas where banks would not provide mortgages because of higher rates of ethnic residents. This example provided is from Philadelphia in 1936. Examine the *Philadelphia Redlining* map, the *Philadelphia* ethnic dot map, and the modern Google Earth™ imagery in conjunction with one another. Look for relationships and discrepancies in the spatial datasets.

4. Select the statement that is not supported by the maps and imagery of Philadelphia.

   A. There are spatial relationships evident between the redlining and dot maps.
   B. The vast majority of the areas identified as "Best" on the redlining map are large single-family homes today.
   C. In general, the areas highlighted as "Hazardous" are higher population density areas.
   D. Today, white and African American residents live in all of the categories identified in the 1936 map.
   E. The areas that have been developed since 1936 are primarily home to white residents today.

Exploration 9.2 SHORT ESSAY

1. After studying the assorted maps of the *Ethnic Cities* folder, which of the included cities combines the highest degree of ethnic diversity for the city as a whole with the highest degree of ethnic separation? Explain your answer.

   _____
   _____
   _____
   _____
   _____
   _____
   _____

2. The term *redlining* has had an evolving meaning over the years. In more recent decades, some have applied the term to situations where banks will loan money to lower income whites before they would lend to lower or even middle and upper income blacks. Think of a neighborhood in your community with an overwhelming presence of one ethnic group. Why do you think this has occurred? Do you believe there is any legacy of redlining in its historic or contemporary meaning that has contributed to this distribution? Provide a few examples to support your opinion.

   _____
   _____
   _____
   _____
   _____
   _____

Exploration 9.3: Ethnic Nationalism

Ethnic nationalism is reflected through the establishment of nation-states, where a largely homogenous ethnic entity exerts control over a defined political territory. True nation-states are very rare in today's multi-ethnic world. On the other hand, there are numerous groups that perceive themselves as ethnic nations without a political state mechanism. Open the *Ethnic Nationalism* folder and then *A Strong Case* folder. In *A Strong Case* folder are a number of ethnic groups that have attempted and/or are continuing to attempt to assert their claim to statehood on the basis of an ethnic identity. Perhaps the strongest case belongs to a group that has a population of more than 20 million, speaks its own language, and has had a continual presence in its region for more than 1,000 years.

Exploration 9.3 MULTIPLE CHOICE

1. Which of the locations placemarked in *A Strong Case* folder is described above?
    A. A
    B. B
    C. C
    D. D
    E. E

South Sudan is the newest state in Africa that has arisen partially as a result of ethnic differences. There are numerous sites/situations that may yet birth more states. Five are placemarked in the *Future African State* folder. Tour the sites and complete outside research to gain a cursory understanding of the situations that are driving the call to independence at each of the locations.

2. Which of the placemarked locations in the *Future African State* folder has the weakest ethnic linkages?
    A. Bioko
    B. Republic of Cabinda
    C. Caprivi Strip
    D. Kabylie
    E. Kingdom of Kongo

Europe has similar issues with a number of secessionist movements with ethnic components. Examine the placemarked sites/situations in the *Future European State* folder. Tour the sites and complete outside research to gain a rudimentary understanding of the secessionist movements at each location.

3. Which of the locations in the *Future European State* folder was involved in a war in 2008?
    A. Aland
    B. Abkhazia
    C. Basque Country
    D. Catalonia
    E. Crimea

In the United States, many Native Americans live in locations that have sovereign status and some level of autonomy from the surrounding political entities. While large reservations such as the Navajo reservation in northeast Arizona may come to mind, many tribes have recently applied their ethnicity-based sovereign status to smaller parcels of land for the economic gain of the tribe. Two examples are found in the *US* folder. Tour the sites and complete outside research to gain a cursory understanding of the primary economic resource for each of the two locations.

4. The sites in the *US* folder are related to
    A. banking.
    B. gaming.
    C. manufacturing.
    D. the military.
    E. telecommunications.

Exploration 9.3 SHORT ESSAY

1. Based on your research for the questions in this exploration, which one of the European or African future states is most likely to gain independence first?

_____
_____
_____
_____
_____
_____
_____

2. Based on your research for the questions in this exploration, describe the ethnic component of one of the European or African future states.

_____
_____
_____
_____
_____
_____
_____

Exploration 9.4: Ethnic Conflict

Unfortunately, conflict with an ethnic bent has been and continues to be a great challenge for humanity. Perhaps the most well-known conflict of recent years has been centered on the African state of Sudan. The westernmost region of Sudan is known as Darfur. Turn on and open the *Crisis in Darfur* folder and explore the content provided by the American Holocaust Memorial Museum under Testimonies.

Exploration 9.4 MULTIPLE CHOICE

1. Based on the information contained in the *Crisis in Darfur* folder, identify the statement that is most strongly supported.

   A. Displaced persons are mainly confined to Western Darfur (Gharn Darfur).
   B. The Janjaweed exempt children and women from their attacks.
   C. Darfur will now be part of the new state of South Sudan.
   D. Rape has been used extensively in Darfur.
   E. Central African Republic has been the main recipient of refugees from Darfur.

In the 1990s, an ethnic flashpoint was the disintegrating remains of Yugoslavia. Sarajevo, the capital city of Bosnia and Herzegovina was under siege from April 1992 to February 1996. Thousands of Serb forces surrounded the city and assaulted it with artillery and sniper fire. This was particularly shocking because the city had been home to a great symbol of peace, the Olympics, in 1984. Olympic venues were destroyed and used for heinous purposes during the siege. While most of the city has since been rebuilt, there are numerous reminders of this ethnic conflict. Some are even visible in Google Earth ™. Go to the *Sarajevo* placemark and view the scene.

2. What is evident from the view associated with the *Sarajevo* placemark?

    A. Giant cemeteries now surround some of the old Olympic venues.
    B. "Sarajevo Roses" are in bloom year round.
    C. The Olympic venues have been repaired, but the housing is still in ruins.
    D. The Olympic facilities have been converted to prisoner of war camps.
    E. The ethnic division of the city is still very clear.

While not of the scope of Darfur or Sarajevo in terms of casualties, the United States has its own twentieth century history of ethnic conflict. Events such as the Rodney King case have periodically caused flares in ethnically oriented conflict. Go to the *Little Rock* placemark and use Google Earth™ and any additional outside resources to educate you as to the significance of this location in terms of ethnic tensions.

3. What term is least directly associated with the *Little Rock* placemark in terms of ethnic conflict?

    A. Little Rock 9
    B. Central High
    C. forced desegregation
    D. Orval Faubus
    E. Edmund Pettus Bridge

A more subtle form of ethnic conflict takes the form of something called environmental racism. This is the notion that ethnic minorities are disproportionately subjected to environmental hazards. Research the concept more using outside sources and then go to the *Altgeld Gardens* placemark. This is a housing project that has been cited in the environmental racism literature.

4. Based on the view from Google Earth™, what can you see that could present a health problem?
    A. A large airport
    B. Evidence of heavy industry
    C. Parks and recreation areas
    D. High crime
    E. Obesity

Exploration 9.4 SHORT ESSAY

1. After exploring the content of the *Crisis in Darfur* folder, you should have an enhanced understanding of the scope of the problem. Do you think that affluent, militarily-capable western countries should have intervened in Darfur? Why do you think they have not?

_____

_____

_____

_____

_____

_____

2. Think about your community. Are there examples of environmental racism? Describe an example.

_____

_____

_____

_____

_____

_____

_____

Encounter Human Geography

Name: _____

Date: _____

Chapter 10: Sexuality and Gender

Sexuality and gender represent social constructions that have increasingly been explored by geographers in recent years. While differentiation between sexes has traditionally been a biological distinction, there are chromosomal patterns that do not fit neatly into male or female categories. Gender is tied more to the concept of roles and identities people associate with males and females. Therefore, one might consider in simplistic terms that sexuality describes who we are physically and gender to what we do.

Download EncounterHG_ch10_Sexuality_and_Gender.kmz from **www.mygeoscienceplace.com** *and open in Google Earth™.*

Exploration 10.1: Health

Health is the first area of exploration that can distinguish divergent paths for men and women around the globe. Reproductive health measures, such as maternal mortality, highlight discrepancies in development and can also illuminate distinct cultural differences. Maternal mortality refers to the death of a woman during or shortly after pregnancy. Causes can include infections, gestational hypertension, unsafe abortions, severe bleeding, and uterine ruptures. Poor maternal nutrition and inadequate medical care often contribute to these causes. Open the *Sexuality and Gender* and the *Health* folders, and turn on the *Maternal Mortality Rate* folder.

Exploration 10.1 MULTIPLE CHOICE

1. Identify the statement regarding the *Maternal Mortality Rate* folder that is most accurate.

   A. Women in less-developed countries are much more likely to die during pregnancy or child birth than women in the more developed countries.
   B. According to this dataset, South Asia is the highest risk location for maternal health.
   C. The Dominican Republic has the highest maternal mortality rates in the Americas.
   D. In 2000, there were only three countries where more than 1 in 1,000 mothers died as a result of pregnancy or child birth.
   E. The United States has the lowest maternal mortality rates in the world.

Turn off the *Maternal Mortality Rate* folder. One contributing factor to the varying maternal mortality rates is the legal status of abortion in respective countries. Where these services are severely restricted or even illegal, more women may die as they give birth in high-risk health situations. Now, activate the *Illegal Abortion* folder. The *Illegal Abortion* folder contains placemarks to five countries with some of the strictest anti-abortion laws. In these countries, women and doctors can be charged with felony offenses for abortion resulting in jail terms as long as ten years.

2. What is the common thread that most likely explains the prohibition on abortion for the countries contained in the *Illegal Abortion* folder?

   A. Sharia law is enforced in these countries.
   B. These are Spanish-speaking countries.
   C. These countries are all World Trade Organization members.
   D. Abortion is illegal throughout the "new world."
   E. These are all countries with a strong Roman Catholic presence.

Inequality among the sexes begins right at birth in many places around the world. For example, turn on the *Sex Ratio at Birth* folder and study the global patterns. While the natural sex ratio is 105:100 in favor of males, many populations have altered this ratio significantly. Oftentimes this is the result of cultural preferences for boys.

3. Which of the following statements is most strongly supported by the *Sex Ratio at Birth* folder?

   A. Young girls do not get the same level of resources dedicated to their upbringing as young boys.
   B. Sex-specific abortion is utilized in some locations around the world.
   C. Europe is generally home to the highest male/female sex ratios at birth.
   D. Sex ratios favoring boys are found only in South Asia and East Asia.
   E. Sex ratios deviate farther from a 1:1 ratio in the most developed countries of the world.

If more males are born around the world than females, then it would follow that there should be higher male populations in most countries. Turn off the *Sex Ratio at Birth* folder and turn on the *Sex Ratio 65 and Older* folder. Does this data support or refute the aforementioned statement?

4. What statement is most strongly supported by *Sex Ratio 65 and Older* folder?

   A. The data suggests that men are stronger than women.
   B. By the age of 65, women outnumber men in all countries of the world.
   C. Russia, through disproportionate male deaths, has approximately two women for every man in the 65 and older age bracket.
   D. The data suggests that over time, women are more likely to adopt unhealthy behaviors like smoking and drinking alcohol excessively.
   E. In the majority of countries, men live longer than women.

Exploration 10.1 SHORT ESSAY

1. The *Illegal Abortion* folder highlighted some of the countries around the world with the most restrictive abortion laws. Find out what countries have the most liberal abortion policies. Is there a regional pattern? Are there common cultural threads?

_____
_____
_____
_____
_____
_____
_____

2. Comment on the particularly low male to female ratio for persons 65 and over in the countries of the former Soviet Union. Provide an explanation of this distinctive regional pattern. Turn off the *Sex Ratio 65 and Older* folder when you have completed this question.

_____
_____
_____
_____
_____
_____

Exploration 10.2: Education

One of the most profound differences between males and females around the globe is found in the area of education. Open the *Education* folder and turn on the *Female Literacy* folder and observe the global pattern. Literacy is loosely defined as the ability to read and write. Needless to say, the opportunities for a person who is literate are far greater than those for a person who is illiterate.

Exploration 10.2 MULTIPLE CHOICE

1. Based on your analysis of the *Female Literacy* folder, what continent displays the lowest levels of female literacy?

   A. Africa
   B. Asia
   C. Europe
   D. North America
   E. South America

The *Female Literacy* folder doesn't tell the entire story because there are some parts of the world that have lower literacy rates for both women and men. A more useful measure can be to compare women's literacy rates as a percentage of men's literacy. This helps to illuminate regions or countries with high discrepancies between educational levels of men and women. Turn off the *Female Literacy* folder and turn on the *Female Literacy as a Percentage of Male Literacy* folder.

2. What statement best summarizes what can be gleaned from the *Female Literacy as a Percentage of Male Literacy* folder?

    A. There are no documented cases where a country has more females that are literate than males.
    B. Globally, women's rates of literacy are substantially lower than men's literacy rates.
    C. This dataset emphasizes literacy in most developed countries.
    D. Southeast Asia is the global leader in women's literacy.
    E. Without exception, women's literacy in Africa trails men's literacy.

Turn off the *Female Literacy as a Percentage of Male Literacy* folder and expand and turn on the *Female Suffrage* folder. The right to vote and/or run for office are described by the term *suffrage*. Female suffrage progressed across the globe beginning in the late nineteenth century. More specifically, the diffusion of women's voting rights progressed in a regional pattern that is modeled by the placemarks in the *Female Suffrage* folder.

3. Research the history of female voting rights and place the countries included in the *Female Suffrage* folder in the order in which women gained the right to vote.

    A. Finland, Canada, Brazil, India, Iran
    B. Brazil, Canada, Finland, India, Iran
    C. Iran, India, Finland, Canada, Brazil
    D. Finland, Iran, Brazil, India, Canada
    E. Canada, Finland, Brazil, Iran, India

While inequity in terms of voting rights has largely been addressed and improved globally, the same cannot be said in terms of economic inequalities. In many countries, females are paid only a fraction of their male counterparts. In the United States, the gender pay gap has continued to decline in recent decades. However, a 19.8 percent gap remains as of 2009, according to the Bureau of Labor Statistics. Additionally, there are significant variations by state. Turn off the *Female Suffrage* folder and view the highest and lowest gender gap states that are identified as "Most Parity: 96.5%" and "Least Parity: 65%," and have been marked with placemarks.

4. Select the most likely explanation for the highest and lowest observations for the so-called "gender gap."

    A. Highest = high percentage in service industry, Lowest = high percentage in tourism
    B. Highest = high percentage of Federal employees, Lowest = high percentage of skilled labor in traditionally male jobs (e.g., oil field worker)
    C. Highest = part-time jobs more likely to be held by females, Lowest = New Orleans has high percentage of female entertainers
    D. Highest = mainly state employees here, Lowest = too many state affirmative action laws
    E. Highest = female hiring quotas, Lowest = low percentage of females in the workforce

Exploration 10.2 SHORT ESSAY

1. Turn on the *Female Legislators and Officials* folder. Is there a spatial relationship between the progression of the female right to vote and women holding political office? Explain your answer.

_____
_____
_____
_____
_____
_____
_____

2. Return to the *Female Literacy as a Percentage of Male Literacy* folder. Think about life as a female in one of the countries with the lowest values. Describe some of the challenges that you might face compared to life for females in your home country.

_____
_____
_____
_____
_____
_____
_____

Exploration 10.3: Roles

The gender roles of men and women often reflect social differences and divergent positions of power. In some parts of the world, such as North Africa and Southwest Asia, the role of women is clearly subordinate to men. One element of gender that women are more strongly associated is prostitution. The legal status of prostitution varies widely around the world. Extremes range from locations where it is legal and regulated to locations where it has been punishable by death. Open the *Roles* folder. The placemarks included in the *Prostitution* folder highlight some of the countries with divergent laws. Use this information and outside resources to answer the following question.

Exploration 10.3 MULTIPLE CHOICE

1. Research the global laws on prostitution and select the statement that is most correct.

   A. Prostitution is illegal in the most developed countries in the world.
   B. The region with the most liberal laws regarding prostitution is North Africa / Southwest Asia.
   C. Less than 30 countries have legal forms of prostitution.
   D. Most European countries have some form of legal prostitution.
   E. Prostitution is illegal in all of the Catholic-oriented countries of South America.

Female genital mutilation (FGM) is a sex-specific form of violence that occurs for cultural and religious reasons in some parts of the world. This refers to procedures that intentionally alter or injure female genital organs for non-medical reasons. The World Health Organization has worked hard to bring an end to the practice because of the lack of consent for the procedure along with concerns about the safety and long-term health consequences. Approximately 100 to 140 million girls and women worldwide are currently living with the consequences of FGM. Go to the World Health Organization Web site and research this situation further.

2. Identify the placemarked location within the *FGM* folder where a young woman would be most likely to suffer an FGM procedure.

    A. A
    B. B
    C. C
    D. D
    E. E

The Gender Inequality Index (GII) reflects the loss in human development due to inequality between female and male achievements. Variables considered are maternal mortality, adolescent fertility, parliamentary representation, educational attainment, and labor force participation. Ten countries have been flagged in the *GII* folder. These represent the five countries with the highest and lowest GII scores globally. Determine what countries should be grouped into each category. Utilize any outside resources necessary.

3. Based on your grouping of the countries into high and low GII, select the statement that is supported.

    A. Africa is home to the top five countries in terms of gender inequality.
    B. Islamic countries are generally leaders in gender equality.
    C. Northern Europe is the leader in gender equality.
    D. The global leaders in gender equality are countries with a variety of predominant religious traditions.
    E. There does not seem to be any spatial relationship regarding gender inequality.

One of the contributing factors to the gender gap in the United States relates to the types of jobs that people have. This can be influenced by male and female dominated fields. Some of the most gendered jobs are illustrated by the placemarked locations in the *Gendered Jobs* folder. The jobs associated with the placemarks are body shop employee, cement mason, dental hygienist, hairdresser, logger, and preschool teacher.

4. Group the placemarks into categories that represent traditional male and female jobs.

    A. Female = 1,2,3; Male = 4,5,6
    B. Female = 3,4,5; Male = 1,2,6
    C. Female = 1,5,6; Male = 2,3,4
    D. Female = 2,3,4; Male = 1,5,6
    E. Female = 4,5,6; Male = 1,2,3

Exploration 10.3 SHORT ESSAY

1. What could explain the spatial pattern evident in terms of high and low GII countries?

_____

_____

_____

_____

_____

_____

2. What are the economic implications of regions where women have vastly inferior rights?

_____

_____

_____

_____

_____

_____

_____

Exploration 10.4: LBGT Geographies

Geographers have increasingly started to study the many ways our cultural identities can vary from what has been deemed "normal" in the past. One way this can happen is to study the way sexual diversity can challenge the notion of the heterosexual norm. There are many diverse representations of sexuality and gender and they are reflected in the spaces and places of the world. Some cities, for example, have become known for the large percentages of LGBT persons. LGBT stands for lesbian, gay, bisexual, and transgender people. Open the _LGBT Geographies_ folder and then open the _LGBT Cities_ folder. This folder contains placemarks for the cities with the highest percentages of persons that identify with the LGBT designation.

Exploration 10.4 MULTIPLE CHOICE

1. Select the statement that is supported by the _LGBT Cities_ folder and any outside research you deem necessary.

   A. The top five cities are all located on the West Coast.
   B. California is home to most of the cities.
   C. The top five cities are in states that have voted for a mixture of Republicans and Democrats at the Presidential level in the last three elections.
   D. These cities are only located in states with very low Catholic populations.
   E. These cities are only located in states with very young populations.

Within urban areas, it is not uncommon to see neighborhoods where higher percentages of LGBT people live. In many cases, the LGBT population has been instrumental in gentrification efforts where urban districts have been revitalized by new investment. Therefore, a number of these gay villages are found in areas with significant historical or architectural resources. Take a look at the selection of street scenes from a variety of gay villages in the _Gay Villages_ folder.

2. Select the statement that is most strongly supported by the _Gay Villages_ folder.

   A. Gay villages are found only in the cities with the highest LGBT population percentages.
   B. One common thread between these sites is the display of rainbow flags and banners.
   C. Gay villages are purely an American phenomenon.
   D. Gay villages are run-down, low-income neighborhoods.
   E. There are no visible cues that the areas in the _Gay Villages_ folder have a LGBT focus.

Go to the _NRHP_ placemark to view a location that is included in the National Register of Historic Places (NRHP). The NRHP is the National Park Service's list of districts, sites, buildings, structures, and objects that have been deemed worthy of historic preservation. There are more than one million sites included on the list. You can go to the NRHP Web site to learn about the different properties on the registry.

3. Based on the National Register of Historic Places, which of the following statements is <u>not</u> correct?

   A. This site is located in Greenwich Village.
   B. The nominated area is known as Stonewall.
   C. The buildings were originally constructed in the 1930s.
   D. The Stonewall event occurred in 1969.
   E. Stonewall is considered by some to be the single most important event that led to the modern gay and lesbian liberation movement.

Same sex marriage is the legally recognized marriage between two people of the same sex. Countries around the world have taken very different approaches to this issue. In some countries, same sex marriage is an offense punishable by death and in others there is little to no distinction made between couples of the same or opposite sex. The *Same Sex Marriage* folder contains four countries with the strictest prohibition on same sex marriage that include the death penalty for homosexuality to four countries that have legalized same sex marriages.

4. Use outside resources to determine the correct grouping of countries in the *Same Sex Marriage* folder.

   A. Same sex marriage legal = 1,2,3,4; Death penalty for homosexuality = 5,6,7,8
   B. Same sex marriage legal = 1,2,3,6; Death penalty for homosexuality = 4,5,7,8
   C. Same sex marriage legal = 4,5,6,7; Death penalty for homosexuality = 1,2,3,8
   D. Same sex marriage legal = 2,3,4,6; Death penalty for homosexuality = 1,5,7,8
   E. Same sex marriage legal = 5,6,7,8; Death penalty for homosexuality = 1,2,3,4

Exploration 10.4 SHORT ESSAY

1. Is there a gay village or LBGT district in your community? If not, is there some other spatial association with the LBGT community?

   _____
   _____
   _____
   _____
   _____
   _____
   _____

2. Hypothesize what the *Same Sex Marriage* map will look like in 100 years.

   _____
   _____
   _____
   _____
   _____
   _____
   _____
   _____

Name: _____

Date: _____

Chapter 11:
Political Geography

Political geography looks at the relationships between politics and space, as represented by human and physical geography. In this chapter, you will explore the concepts of states, boundaries, the political landscape, and geopolitics. It is an interesting time to study these facets of political geography, as the world simultaneously undergoes convergence through greater international political cooperation and divergence as local cultural groups demand more autonomy.

*Download EncounterHG_ch11_Political_Geography.kmz from **www.mygeoscienceplace.com** and open in Google Earth™.*

Exploration 11.1: States

The World Factbook, published by the Central Intelligence Agency, is one of the best sources for readily accessible information on state entities. Topics include history, people, government, economy, geography, communications, transportation, military, and transnational issues. Additionally, comparisons can be made among countries in a variety of statistical measures. Open the *States* folder and you will see a link to the Factbook associated with the *CIA World Factbook* folder. Click this link. Also, inside there is another folder titled *Central America*. Open this folder and view the placemarked countries. In addition, make sure you have activated the *Borders and Labels* folder in Google Earth™.

Exploration 11.2 MULTIPLE CHOICE

1. Use the information available from the Factbook to identify the placemarked state in the *Central America* folder that matches the following criteria: democratic republic, population between four and five million, and slightly smaller than the state of West Virginia in terms of area.

    A. A
    B. B
    C. C
    D. D
    E. E

Turn off the *Central America* folder and open and turn on the *Enclaves/Exclaves* folder.

Enclaves and/or exclaves create unique challenges for state governments. While enclaves are completely surrounded by foreign territory, an exclave is attached politically but not physically to another territory. Enclaves can also be exclaves and vice versa, but not always. There are a number of situations around the world where this complicated political geography can be the source of international tension. This is a good example of the type of information one can find in the *Factbook*.

2. Utilizing information regarding the transnational issues of countries found in the *Factbook* country profiles, identify the placemarked state that has challenges associated with enclaves and/or exclaves.

    A. Armenia
    B. Chad
    C. Ghana
    D. Russia
    E. South Africa

Turn off the *Enclaves/Exclaves* folder. A buffer state is a country that is situated between two more powerful states that are potentially hostile to one another. A true buffer state is independent with a neutral foreign policy that favors neither of the adjacent states. A historical example is Belgium, which served as a buffer state between France, the German Empire, and the United Kingdom prior to World War I. Open and turn on the *Buffer State* folder.

3. From the selection of states included in the *Buffer State* folder, what's the best example of a buffer state in today's world?

    A. A
    B. B
    C. C
    D. D
    E. E

Turn off the *Buffer State* folder and turn on the *Visible Boundary* folder. State boundaries can range from conspicuous to invisible depending upon the environmental, economic, and political factors at a given location. Beyond the actual boundary that may or may not be delimited by a fence, one can occasionally see stark differences from one state to the next.

4. In the case of the *Visible Boundary* folder, examine the placemarked border regions and identify the location that exhibits the most striking difference in the physical environment from one side of the border to the other.

    A. Peru/Bolivia
    B. Vietnam/Laos
    C. India/Pakistan
    D. Haiti/Dominican Republic
    E. Tanzania/Mozambique

Exploration 11.1 SHORT ESSAY

1. You identified a state boundary with visible differences in the physical environment above. Now look for a state boundary where distinct differences can be noted in the human environment. Perhaps one side of the boundary is densely developed and the other is not. Describe the location and the differences you observe.

_____

_____

_____

_____

_____

_____

_____

2. Go to the *Kosovo* placemark. Now search the Web for more information about Kosovo. Why are some of the borders of the state red in Google Earth™? What is the current status of Kosovo according to Serbia?

_____

_____

_____

_____

_____

_____

_____

Exploration 11.2: Boundaries

The distribution of national territory is a key geographical aspect of the state. There are significant benefits to states that are compact. These include less border to defend, shorter communication lines, and less transportation infrastructure to maintain. Elongated, prorupted, perforated, and fragmented states, on the other hand, all face unique challenges. Open the *Boundaries* folder and then open and turn on the *State Shapes* folder.

Exploration 11.2 MULTIPLE CHOICE

1. Use Google Earth™ to match the states with the most appropriate shape description.

   A. 1 = Compact, 2 = Elongated, 3 = Prorupted, 4 = Perforated, 5 = Fragmented
   B. 1 = Fragmented, 2 = Perforated, 3 = Prorupted, 4 = Elongated, 5 = Compact
   C. 1 = Elongated, 2 = Fragmented, 3 = Compact, 4 = Prorupted, 5 = Perforated
   D. 1 = Prorupted, 2 = Compact, 3 = Elongated, 4 = Perforated, 5 = Fragmented
   E. 1 = Fragmented, 2 = Prorupted, 3 = Perforated, 4 = Compact, 5 = Elongated

Turn off the *State Shapes* folder. States' shapes are defined by their boundaries. There are a wide variety of boundaries ranging from physical boundaries made up of mountains, deserts, or water features to cultural boundaries that are geometric, religious, or language-based.

2. Go to the placemarked boundaries included in the *State Boundaries* folder and identify the most appropriate descriptive term for each.

    A. 1 = language, 2 = desert, 3 = religious, 4 = water, 5 = mountain
    B. 1 = geometric, 2 = water, 3 = religious, 4 = desert, 5 = mountain
    C. 1 = geometric, 2 = water, 3 = desert, 4 = geometric, 5 = mountain
    D. 1 = mountain, 2 = water, 3 = language, 4 = desert, 5 = geometric
    E. 1 = water, 2 = geometric, 3 = mountain, 4 = desert, 5 = language

State boundaries have been and continue to be the source of disputes around the globe. In Google Earth™, disputed boundaries are represented by red boundaries.

3. Which of the following countries does not have a disputed boundary?

    A. Bhutan
    B. China
    C. Pakistan
    D. Qatar
    E. Tajikistan

There are a number of boundaries that are found within the same country. These nested boundaries include things like state, county, and city boundaries along with political districts. Following each census, new boundaries are drafted for congressional districts. These reflect the changes that have occurred in the population. When redistricting is left to the political party in control, the boundaries are usually drawn to maximize the results for that party. Various strategies are employed to concentrate opposition voters into a few districts or spread them so thin across many districts that they are unable to win any seats. The practice is called gerrymandering and can create some uniquely shaped congressional districts. Turn on the *Texas Congressional Districts* folder to see the breakdown for the 111th Congress.

4. Select the statement that is supported by the *Texas Congressional Districts* folder.

    A. The largest district is the 19th.
    B. The largest cities (e.g., Houston) are squeezed into a maximum of two districts.
    C. Congressional districts and counties never share the same boundaries.
    D. The larger districts are in the western half of the state.
    E. Amarillo and Lubbock are in the same district.

Exploration 11.2 SHORT ESSAY

1. Think about where you live. What are the nested political boundaries that surround you?

_____
_____
_____
_____
_____
_____
_____

2. What types of boundaries are utilized to delimit your state? Why does your state have its shape?

_____

_____

_____

_____

_____

_____

_____

Exploration 11.3: The Political Landscape

Relic boundaries are borders that no longer exist as international boundaries but still leave a cultural imprint nonetheless. Old international borders may act as new sub-state borders, for example. Other times, economic or social divides may exist on opposite sides of the retired boundary. In some cases, the border may be memorialized as a political symbol. Open the *Political Landscape* folder and then open the *Relic Boundaries* folder to view a sample of relic boundaries. Each placemark has a hint word or phrase associated with the placemark.

Exploration 11.3 MULTIPLE CHOICE

1. Identify the placemarked location in the *Relic Boundary* folder that is least appropriately classified as a relic boundary.

   A. A
   B. B
   C. C
   D. D
   E. E

There are a number of additional boundaries within states that are not readily visible. One set is the Public Land Survey (PLS). The system divides much of the central and western United States into a rectangular system of surveys used for legal land descriptions. With the advent of the Global Positioning System (GPS), the PLS is not used as extensively for location purposes. Open the *Public Land Survey* folder to access a link to an article on the PLS. Turn on the *PLS Overlay* to see a map of the principal meridians and base lines of the United States. Unfortunately, the overlay's projection does not facilitate a very good match with Google Earth™, so you may find it helpful to turn off the *Borders and Labels* folder in the Primary Database.

2. Using information from the PLS article and the *PLS Overlay*, what is the name of the principal meridian for the location of Mount Rushmore?

   A. Black Hills
   B. Wyoming
   C. Wind River
   D. 6th
   E. South Dakota

State capitol buildings are but one way the power of the state or sub-state entity is displayed. These capitol buildings are important focal points of the state political system. Usually the state's governor is based in this building along with state legislatures and sometimes state supreme courts. Turn off the *PLS Overlay*. View the sample of US state capitols in the *Capitol Buildings* folder. Notice the prevalence of "old world" architectural styles that mimic the federal capitol building in Washington, DC. Two more examples of "old world" architecture can be viewed by flying to St. Peter's Cathedral at the Vatican or the Parthenon in Athens.

3. What state's capitol does not make widespread use of Greek and/or Roman architectural motifs?

    A. Alabama
    B. California
    C. Kentucky
    D. New Mexico
    E. Utah

Turn on the *Borders and Labels* folder in the Primary Database. While a capitol building presents a rather stark reminder of the government, we are often surrounded by a variety of features that also bare testament to the political power of the state. The *State Works* folder provides a few examples of the ubiquitous presence of the state through its public works projects and financing.

4. Identify the political landscape element that least effectively demonstrates the impacts of political authority and funding.

    A. A
    B. B
    C. C
    D. D
    E. E

Exploration 11.3 SHORT ESSAY

1. What is the Public Land Survey description of the building that houses your classroom for this class? Provide your response to the nearest quarter section. Also name the principal meridian for your area if it is not a generic state principal meridian. If you live in an area without the PLS, note it and find the PLS description for the state capitol building in New Mexico. Hint: Use the National Atlas (*www.nationalatlas.gov*) or National Map (*www.nationalmap.gov*) to help you. Both sites enable you to search using the PLS and also provide detailed text explaining why the system was established and how it works.

_____
_____
_____
_____
_____
_____
_____

2. Mount Rushmore, mentioned above, is a good example of an expression of American nationalism. Think about who is responsible for creating this component of the political landscape. These places are created by people with access to power and resources to memorialize important symbols to them. Do you think the native Lakota Sioux value this site in the same way as European Americans? Answer this question and then fly to the *Alternative Icon* placemark and turn on the *3D Buildings* layer. What is this site and what is its purpose?

_____

_____

_____

_____

_____

_____

_____

Exploration 11.4: Geopolitics

A nation is a group of people sharing a common cultural trait, history, or political identity. Many times, a nation is contained within a state. In the case of the Soviet Union, there were many nations that were incorporated into the Russian state as it expanded over a 500 year period. In 1991, the Soviet Union dissolved and the peripheral nations that had been incorporated into the Soviet Union regained their independence. Open the *Geopolitics* folder and then open and turn on the *Soviet Union* folder.

Exploration 11.4 MULTIPLE CHOICE

1. Which of the placemarked locations was not part of the Soviet Union?

    A. A
    B. B
    C. C
    D. D
    E. E

Turn off the *Soviet Union* folder. In the case of the Soviet Union, there were great centrifugal forces that worked to pull apart the state. The Soviet state therefore used a variety of techniques to create centripetal forces that were intended to keep the Soviet Union together. Most were unsuccessful, as Russian culture was forced upon the very diverse national identities of the peripheral locations. View the locations placemarked in the *Centrifugal Forces* folder.

2. Identify the location that is best matched with a centrifugal force that would have pushed it away from the Soviet Union.

    A. A = Judaism
    B. B = Catholicism
    C. C = Altaic language
    D. D = Islam
    E. E = Uralic language

Turn off the *Centrifugal Forces* folder. While the Soviet Union dissolved under the weight of its centrifugal forces, there are a growing number of collaborations and partnerships among states. Many of these supranational organizations have political and economic goals that dictate diminished state sovereignty. The European Union (EU) is a good example of a supranational organization. The EU continues to be a dynamic entity with evolving goals and an ongoing desire for expansion by some countries not yet included into the EU. Another facet of the EU is the common currency employed by the majority of member states. The Euro has grown to the second largest reserve currency in the world as of 2010. Some EU members have resisted adopting the common currency as they seek to maintain more independent control over their monetary policies.

3. Identify the states in the *Euro* folder that are placemarked with the Euro symbol that are <u>not</u> a part of the European Union's Eurozone.

    A. Sweden, United Kingdom
    B. Bulgaria, Denmark
    C. Malta, Greece
    D. Portugal, Greece
    E. Spain, Netherlands

Seven countries have claimed extensive portions of Antarctica, and several of these claims overlap. However, the world at large does not recognize claims on Antarctica as the continent is considered politically neutral. All kinds of important scientific research takes place at a host of research stations scattered around the continent. Five of the research stations in Antarctica have been placemarked in the *Antarctica* folder. Utilize the *Photos* layer and any outside resources to gain insight into what states operate the research stations in Antarctica.

4. What are the countries represented by the flagged Antarctic research stations?

    A. Germany, India, Japan, Poland, United States
    B. Australia, France, New Zealand, United Kingdom, United States
    C. Argentina, Chile, France, Norway, United Kingdom
    D. Argentina, Australia, Chile, France, Norway
    E. China, New Zealand, Russia, South Africa, United States

Exploration 11.4 SHORT ESSAY

1. EU expansion is a popular topic in the region. Research the issue and determine a country that is most likely to be a candidate for the next expansion. Identify several strengths and weaknesses regarding the bid for membership.

    _____
    _____
    _____
    _____
    _____
    _____
    _____

2. What is the relationship between the claimed regions of Antarctica and the distribution and national origin of research stations?

_____
_____
_____
_____
_____
_____
_____

Name: _____

Date: _____

Chapter 12:
Conflict

Conflict has proven to be an inherent part of the human condition. The geography of conflict is diverse across space and time, and ranges from local to global. Some conflicts have persisted for millennia, while others have proven to be ephemeral disruptions. This chapter introduces you to some of these myriad places of conflict. Additionally, the resources that can spur conflicts are explored. The subsequent development of military landscapes and the contemporary challenges of terrorism are also examined.

Download EncounterHG_ch12_Conflict.kmz from **www.mygeoscienceplace.com** *and open in Google Earth™.*

Exploration 12.1: Places of Conflict

Open the *Places of Conflict* folder followed by the *Ongoing Conflicts* folder. The placemarked locations represent places of ongoing conflict around the globe. Some of these conflicts have been ongoing for decades, while others have arisen in the last few years. The scale of these conflicts is also divergent, ranging from several hundred to many thousands of combatants at any time. Research each site utilizing Google Earth™ resources and any additional materials you choose.

Exploration 12.1 MULTIPLE CHOICE

1. Match the clues with the placemarked locations in the *Ongoing Conflicts* folder.

    A. 1 = Shiite insurgency; 2 = Dates to 1978; 3 = Drug war; 4 = Self-determination movement; 5 = 2007 Emirate declaration
    B. 1 = Dates to 1978; 2 = Drug war; 3 = Shiite insurgency; 4 = 2007 Emirate declaration; 5 = Self-determination movement
    C. 1 = Drug war; 2 = Self-determination movement; 3 = 2007 Emirate declaration; 4 = Shiite insurgency; 5 = Dates to 1978
    D. 1 = Self-determination movement; 2 = 2007 Emirate declaration; 3 = Shiite insurgency; 4 = Dates to 1978; 5 = Drug war
    E. 1 = Dates to 1978; 2 = Shiite insurgency; 3 = Drug war; 4 = Self-determination movement; 5 = 2007 Emirate declaration

Go to the *Battle of Mogadishu* placemark in the country of Somalia. This site marks the focus of a famous battle that occurred here in 1993. You may have seen the popular film *Blackhawk*

Down, which was based on the events of this conflict.

2. What characteristic of the surroundings near the *Battle of Mogadishu* placemark would have provided the greatest challenges for soldiers on the ground?

    A. The skyscrapers providing sniper roosts.
    B. The open spaces that allow for extensive mine fields.
    C. The tightly spaced and irregular urban environment.
    D. The areas of high ground provided by the hilly topography.
    E. The streams that act as barriers to troop movement.

Somalia is ranked as the world's number one failed state according to *Foreign Policy* magazine. A failed state is one that has failed at meeting some or all of the expectations associated with a sovereign state. This can range from loss of control of its territory to a failure to provide basic services. Not surprisingly, failed states are often home to conflict. Open the link to Foreign Policy's Failed States Index.

3. According to the 2010 rankings of failed states, which of the following statements is most strongly supported?

    A. The majority of states in the top 10 are located in Asia.
    B. Somalia and Sudan have the poorest human rights scores on Earth.
    C. The greatest risks associated with Bosnia Herzegovina, the only European state in the top 100, stem from demographic pressures.
    D. Colombia is the state with the highest-risk for failure in the western hemisphere.
    E. The most problematic aspect for the United States in the Failed State Index is the de-legitimization of the state.

Go to the *Nagasaki* placemark. Nagasaki was one of two cities where nuclear bombs were dropped near the conclusion of World War II. Open the link associated with the *Nagasaki* placemark. This Nagasaki Archive is an interactive Google Earth™ project that allows you to see map overlays and read interviews with persons who were located near the blast zone. The checkboxes at the top of the page enable you to turn on and off the different features. The English language accounts are represented by the Hibakusha (E) features. Explore this project. Be sure to examine some of the historic photographs, as well as the contemporary *Photos* layer. If you have trouble with this application running smoothly, you should open it independently in your Web browser.

4. What is at the site of Nagasaki's Ground Zero today?

    A. Museum of History and Culture
    B. A fountain
    C. A baseball stadium
    D. A black pillar
    E. The A-Bomb Dome

Exploration 12.1 SHORT ESSAY

1. Identify a state other than Somalia that is ranked in the top 10 in the Failed State Index in 2010. Describe some of the unique pressures facing that state's efforts at maintaining a viable state.

_____
_____
_____
_____
_____
_____

2. After examining the images, interviews, and maps associated with the Nagasaki atomic bomb, describe your enhanced geographic understanding of the event and site. For example, was it bigger or smaller than you had envisioned? Were you aware of the immediate effects on humans? Does the modern urban scene meet your expectations? Add anything else that demonstrates your new knowledge.

_____
_____
_____
_____
_____
_____

Exploration 12.2: Resource Conflict

In the Niger Delta, there has been an ongoing low-level guerilla war since 2004. Open the *Resource Conflict* folder and click on the *Niger Delta* placemark. As you explore the delta region, you can see distinctive features associated with this type of resource extraction. Use this information and any information you gain from outside research to learn about this conflict. A good place to start is to look up information on environmental activist Ken Saro-Wiwa.

Exploration 12.2 MULTIPLE CHOICE

1. What is the principal resource extracted in the Niger Delta?

   A. water
   B. oil
   C. hardwood forests
   D. diamonds
   E. gold

Many people have said the wars of the future will be fought over water. Hundreds of millions of people around the world lack sufficient access to water for household and agricultural purposes. The problems are most acute in semi-arid and arid regions of the world. The Euphrates and Tigris Rivers of Southwest Asia represent one river system that has been dammed extensively to provide water for irrigation in the region along with hydroelectricity. The Ataturk Dam is the largest of the more than 20 dams on the system. The potential for political conflict increases dramatically when dams are built on waterways that flow through multiple states.

2. Go to the *Ataturk Dam* placemark and determine what states are located downstream of the dam.

    A. Turkey, Syria, Iraq
    B. Syria, Iraq, Lebanon
    C. Iran, Lebanon, Turkey
    D. Syria, Turkey, Bulgaria
    E. Turkey, Syria, Jordan

There are a number of resource-rich disputed territories around the world that are a source for potential conflict. In a world of finite resources, states continue to press for the maximization of their resource base. Each of the placemarks included in the *Disputed Territory* folder represents a location where there is an ongoing territorial dispute.

3. Identify the statement regarding the placemarks included in the *Disputed Territory* folder that is most strongly supported by Google Earth™ and your outside research.

    A. The northern islands of this chain are disputed between Russia and Japan.
    B. Oil and gas disputed between Philippines, Vietnam, Taiwan, Malaysia, and China.
    C. Mauritius wants access to oil and gas.
    D. Malaysia, Brunei, and Indonesia argue over fishing rights here.
    E. Brazil claims mineral and fishing rights violated by United Kingdom here.

Conflict does not need to involve armies of opposing sides to have enormous impacts on affected states. Take, for example, the *Transboundary* placemark. This site is associated with a major international row between two countries that are normally close allies. In the 1990s and early 2000s, the countries were able to address the problem with a series of international agreements. Research the location to determine the environmental significance of the site.

4. What answer best describes the issue associated with the *Transboundary* placemark?

    A. Greenland suing the United States for damages from global warming.
    B. Mexico unhappy with NAFTA-related regulations on nuclear power.
    C. Acid rain in the United States from Canadian pollution.
    D. Nuclear pollution in Canada from the United States.
    E. Acid rain in Canada from US pollution.

Exploration 12.2 SHORT ESSAY

1. Think of another transboundary resource issue that has led to contention between state (countries) or sub-state (e.g., US states, provinces, municipalities) actors?

    _____
    _____
    _____
    _____
    _____
    _____
    _____

2. Go to the *Arctic Resources* placemark. What dominant physical characteristic of this cold region is not shown by Google Earth™? Why? What does this have to do with resources? Why is this increasingly relevant?

_____

_____

_____

_____

_____

_____

_____

Exploration 12.3: Military Landscapes

The Victory Base Complex is a large military installation that was built at the Baghdad International Airport after the US-led invasion of Iraq. Zooming in to the imagery allows you to see the extensive outlay of housing and military equipment assembled at the site. Use the historical imagery time slider to see the evolution over time.

Exploration 12.3 MULTIPLE CHOICE

1. Based <u>solely</u> on available imagery in Google Earth™, when does the US invasion take place?

   A. Prior to 3/31/2002
   B. Between 6/29/2002 and 2/08/2003
   C. 3/20/2003
   D. Between 9/25/2003 and 6/03/2004
   E. After 1/05/2005

The Victory Base Complex is a specialized base that was established in response to a unique geopolitical situation in Iraq. There is a variety of other bases with specialized missions scattered around the globe. Many bases make use of distinctive geographic features or add their own unique built forms to the cultural landscape. Open the *Bases* folder to see several of these placemarked bases.

2. Match the numbered placemarks with the following descriptive terms for four of the sites: Intercontinental Ballistic Missile warnings; chemical and biological weapons testing; ionospheric disruption; magnetic silencing.

   A. 7;3;1;8
   B. 2;6;4;5
   C. 9;4;6;7
   D. 1;7;3;2
   E. 5;2;7;9

Fortified borders create hard barriers between two political entities. Fortified borders are constructed with stone, earth, bricks and fences. Fortified borders can strengthen the power one group has over another by creating barriers between different cultures, classes, races, and ethnicities. These types of boundaries have existed for thousands of years and continue to be employed to this day. Open the *Fortified Border* folder and view the placemarked locations.

3. Utilizing your knowledge and any additional outside resources, order the fortified borders into a chronological sequence from oldest to youngest according to when they were built.

   A. 1;2;3;4;5
   B. 5;1;2;3;4
   C. 4;5;1;2;3
   D. 3;4;5;1;2
   E. 2;3;5;1;4

Scorched earth is a military strategy that involves destroying resources that are likely to be of use to an advancing army or a society that seeks to rebuild. Is has been used from ancient through modern times in conflicts around the globe. Open the *Scorched Earth* folder to view five example sites.

4. Research the locations in the *Scorched Earth* folder to determine the appropriate match for the dates/sites and key phrases.

   A. 1810 = retreating forces destroy food sources; 1864a = destroying industry and infrastructure; 1864b = destruction of livestock to force starvation; 1938 = flood created to slow advancing troops
   B. 1810 = flood created to slow advancing troops; 1864a = destroying industry and infrastructure; 1864b = destruction of livestock to force starvation; 1938 = retreating forces destroy food sources
   C. 1810 = flood created to slow advancing troops; 1864a = destruction of livestock to force starvation; 1864b = destroying industry and infrastructure; 1938 = retreating forces destroy food sources
   D. 1810 = destroying industry and infrastructure; 1864a = destruction of livestock to force starvation; 1864b = flood created to slow advancing troops; 1938 = retreating forces destroy food sources
   E. 1810 = destruction of livestock to force starvation; 1864a = destroying industry and infrastructure; 1864b = flood created to slow advancing troops; 1938 = retreating forces destroy food sources

Exploration 12.3 SHORT ESSAY

1. Select a specialized base from the *Bases* folder and complete the necessary research to describe the geographic significance of its location.

   _____
   _____
   _____
   _____
   _____
   _____
   _____

2. Select a fortified border from the *Fortified Borders* folder and complete the necessary research to elaborate on its use/necessity

_____

_____

_____

_____

_____

_____

_____

Exploration 12.4: Terrorism

Terrorism has been a part of conflict for centuries. Terrorist acts are violent acts perpetrated for a religious, political, or ideological goal. Most citizens of the United States had not felt the direct or indirect effects of terrorism prior to the September 11, 2001 attacks. While the World Trade Center was the location of the greatest loss of life, two other sites were impacted. One of the planes crashed in a field in Pennsylvania and the other hit its intended target, the US Pentagon. Locate the Pentagon in Google Earth™ and then use the historical imagery capabilities of Google Earth™ to evaluate the scene.

Exploration 12.4 MULTIPLE CHOICE

1. Which of the following statements regarding the Pentagon and the 9/11 attack of the site is best supported by Google Earth™?

   A. The 9/11 Memorial is located on the north side of the Pentagon.
   B. The plane impacted the southeast corner of the Pentagon.
   C. The plane destroyed the first four corridors, but the fifth remained intact.
   D. The 9/11 Memorial appears to be substantially completed by mid 2010.
   E. It appears reconstruction of the Pentagon was complete less than a year after the site was impacted.

Domestic terrorism had struck the United States in April of 1995 when a truck bomb killed 168 people in downtown Oklahoma City at the Federal Courthouse. Timothy McVeigh, a sympathizer of the American militia movement, detonated the bomb. Today, a memorial at the site honors the victims, survivors, and rescuers of the Oklahoma City bombing. Research the symbolic landscape of the Oklahoma City National Memorial by going to the Oklahoma City National Memorial and Museum Web site.

2. Identify the statement that best describes a symbolic element of the Oklahoma City National Memorial.

   A. The Gates of Time frame the moment of destruction.
   B. The Field of Empty Chairs represents the first responders to the scene.
   C. The Survivor Tree was planted after the attack to honor those that survived.
   D. The Orchard represents all the children that were lost in the attack.
   E. The Freedom Tower represents rebuilding that has taken place at the site of the attack and is symbolic of hope.

The *Animal Liberation Front* (ALF) placemark documents a location where ALF performed a criminal activity in the name of animal rights. Study the historic Google Earth™ imagery and research this location to determine what happened here.

3. Select the statement that summarizes the criminal activity performed by ALF at the placemarked location.

    A. Burned a cockfighting arena.
    B. Attacked crocodile hunters' vehicles by slashing their tires and breaking windows.
    C. Released monkeys undergoing testing for cosmetics.
    D. Burned a forest where endangered birds were being captured for their colorful feathers.
    E. Released the laboratory rats of Louisiana State University's biology program.

Now go to the *Earth Liberation Front* placemark. The Earth Liberation Front (ELF) seeks to stop activities that harm the environment. Actions of ELF have been classified as terrorist activities by the US government. This placemark documents the location of one of the higher profile operations of the organization. Research what happened at the *Earth Liberation Front* placemark.

4. Based on your research of ELF and the Google Earth™ imagery, identify the statement that is supported.

    A. Two Elks Lodge was not rebuilt.
    B. ELF protested the operation Breckenridge ski resort at this site.
    C. The Colorado moose population was a concern of ELF at this location.
    D. ELF caused millions of dollars of damage at this site.
    E. The 2005 imagery of the site clearly shows the impacts of ELF's activities.

Exploration 12.2 SHORT ESSAY

1. Go to the *9/11* placemark. Use the historical imagery capabilities of Google Earth™ to provide a brief timeline of the 9/11 site from the earliest available imagery up to the most current imagery.

    _____
    _____
    _____
    _____
    _____
    _____

2. Do you agree with the classification of groups such as ALF and ELF as terrorist groups based on the causes they represent and the actions they have done in the name of their causes?

    _____
    _____
    _____
    _____
    _____
    _____
    _____

Name: _____

Date: _____

Chapter 13: Development

Development is a key theme of geographic study in the twenty-first century. Geographers seek to illuminate and understand the distinct differences in the levels of material conditions between and within the world's regions and countries. This exploration introduces you to some of the elements that contribute to parts of the world being relatively more or less developed. Special attention is paid to the role of the environment in development and the disparity of development within the United States.

Download EncounterHG_ch13_Development.kmz from *www.mygeoscienceplace.com* and open in Google Earth™.

Exploration 13.1: Contributors to Development

The wide range in levels of development around the world is a complex product of a number of factors. These include economic, social, and demographic components. In the economic realm, gross domestic product (GDP) per capita is the most commonly used measure of development. GDP per capita represents the value of all the goods and services produced in a country during a year divided by the number of people that live in the country. In other words, this is a rough and quick measure of affluence around the globe. Open the *Development* folder, then open the *Contributors to Development* folder, and then turn on the *GDP per Capita* folder. Study the global patterns of GDP per capita.

Exploration 13.1 MULTIPLE CHOICE

1. Which of the following statements is best supported by the *GDP per Capita* folder?

    A. Russia was a top 10 country in GDP per Capita in 2008.
    B. South Africa had the highest GDP per capita in Africa in 2008.
    C. In 2008, Luxembourg had the highest GDP per Capita in Europe.
    D. The United States had the highest GDP per Capita in the world in 2008.
    E. As a region, South Asia was the second most affluent in the world.

The Gini index measures the degree of inequality in the distribution of income in a country. The lower the Gini index, the more nearly equal a country's income distribution. Conversely, the higher the Gini index, the more unequal a country's income distribution. In other words, a country's wealth is concentrated in the hands of only a small percentage of people. Turn off the *GDP per Capita* folder and turn on the *Gini Index* folder to see the global variation in income distribution by country.

2. Which of the following statements is best supported by the *Gini Index* folder?

   A. There is more income inequality in Canada than the United States.
   B. The region with the highest degree of inequality is Central America.
   C. China has a greater degree of income equality than India.
   D. Southern Africa has more income inequality than Western Africa.
   E. Portugal exhibits the highest degree of income equality on the Iberian Peninsula.

3. Utilizing the *Gini Index* and *GDP per Capita* folders, what country exhibits the best combination of high income and equal income distribution?

   A. Bangladesh
   B. Jordan
   C. Norway
   D. Peru
   E. Zimbabwe

Turn off the *Gini Index* and *GDP per Capita* folders. Two social indicators of development include education and health care. Evaluating the presence of facilities that are reflective of these indicators can provide a measure of the level of development. Go to the *Luanda* and *Boston* placemarks. Use the "Find Businesses" tab in the Search pane of Google Earth™ to compare the distribution and frequency of relevant sites. Use the search terms "Hospital" and "University." Keep in mind there may be additional pages of search results beyond the initial results displayed. Additional search results are indicated by numbers at the bottom of the search pane.

4. Compare the search results from Luanda, Angola, and Boston, Massachusetts, regarding higher learning and health care and select the most appropriate response.

   A. Luanda has more health facilities while Boston has more higher education facilities.
   B. Luanda has more higher education facilities while Boston has more health facilities.
   C. Boston has more of both health facilities and higher education facilities.
   D. Luanda has more of both health facilities and higher education facilities.
   E. There are roughly similar distributions of higher education and health facilities in both cities.

Exploration 13.1 SHORT ESSAY

1. Write a short paragraph that describes how you can tell if a given country has attained the highest level of development. Include multiple indicators of development in your response.

_____

_____

_____

_____

_____

_____

_____

_____

2. In your opinion, what is the single most important indicator of a high level of development? For example, you could select literacy rate or gross domestic product per capita. Explain your response.

_____

_____

_____

_____

_____

_____

_____

Exploration 13.2: The Development Continuum

There is a remarkable range of development found across the globe in the twenty-first century. While we can make broad generalizations about development regionally or by country, in reality there are great discrepancies to be found within each region and within each country. Open the _Development Continuum_ folder and the _Development Views_ folder. Disregard the country where these placemarks are located. Concentrate solely on the characteristics of what is visible at the placemarked location (e.g., modern infrastructure and technology).

Exploration 13.2 MULTIPLE CHOICE

1. In the three _Development Views_, which best represent more developed and less developed scenes?

   A. 1 = more developed; 2 and 3 less developed
   B. 2 = more developed; 1 and 3 less developed
   C. 2 and 3 more developed; 1 less developed
   D. 1 and 2 more developed; 3 less developed
   E. 1 and 3 more developed; 2 less developed

In spite of the internal variation in development within regions and countries, it can be useful to summarize development characteristics in order to categorize regions. Simply, regions of development can be divided into "haves" and "have-nots." Examine the five proposed divisions of the globe into more and less developed halves contained in the *Global Division* folder. If necessary, review the *GDP per Capita* folder.

2. Within the *Global Division* folder, identify the most appropriate division of the globe into more and less developed halves. Disregard the development status of Australia and New Zealand for this question.
   A. A
   B. B
   C. C
   D. D
   E. E

Turn off the *Global Division* folder and open and turn on the *Very High HDI* folder. Nine of the ten highest ranked countries according to the 2010 United Nations Human Development Index (HDI) have been placemarked.

3. Based on the distribution of highly ranked HDI countries and your knowledge GDP per Capita, which of the following is most likely to be a top ten HDI country?
   A. Barbados
   B. Tonga
   C. Uruguay
   D. Netherlands
   E. Tunisia

Close and turn off the *HDI folder* and open the *Out of Place Scene* folder. Use the linked UN HDI data as you view the placemarked scenes contained in this folder. Determine the scene that seems most out of place with respect to the country's HDI ranking.

4. Which scene is least congruous (most out of place) with regard to the country's HDI ranking?
   A. A
   B. B
   C. C
   D. D
   E. E

Exploration 13.2 SHORT ESSAY

1. In your opinion, what country do you think possesses strong opportunities for HDI improvement in the coming decade? Explain your answer and utilize United Nations HDI data in your response.

_____

_____

_____

_____

_____

_____

_____

2. Provide a brief explanation of the single factor that you think best explains the striking regional divide between more developed and less developed countries that you identified in the *Global Division* question above.

_____

_____

_____

_____

_____

_____

_____

Exploration 13.3: Location and Development

The Earth's varied natural environments present a range of challenges and opportunities for the humans who inhabit these ecosystems. The physical landscape is one of many elements that strongly influence the opportunities for human settlement and development. Imagine you are going to start a society from the ground up. You anticipate that the population of your society will quickly reach 100,000 persons. You will need access to basic resources such as wood and water and have relatively level ground for farming. Now, within the *Location and Development* folder, activate the *New Society* folder and view the placemarks.

Exploration 13.3 MULTIPLE CHOICE

1. Based on the physical geography and what you can see at each placemarked view, which placemarked site included in the *New Society* folder presents the best opportunities for success?

   A. A
   B. B
   C. C
   D. D
   E. E

Turn off the *New Society* folder. The environment also provides daunting challenges in terms of maintaining the health of local populations as certain diseases thrive in particular climates. Subsequently, the burden of coping with these diseases can present a roadblock to higher levels of development. Open the *Infectious Diseases* folder and view the placemarked countries. Now, click on the link that will take you to the CIA's *World Factbook* to identify major infectious diseases that affect each placemarked country. The information is found in the "People" section in each country profile.

2. Identify the location that has significant health risks associated with malaria, schistosomiasis, and typhoid fever.

   A. Argentina
   B. Sri Lanka
   C. Denmark
   D. Haiti
   E. Central African Republic

Turn off the *Infectious Diseases* folder and turn on the *Best Transportation* folder. Another factor that can guide development is proximity to transportation. If a region or city has a good transportation infrastructure, such as highways, railroads, airports, and port facilities, the opportunities for economic growth and subsequent development multiply. Oftentimes, the degree of transportation that exists is a direct reflection of the physical environment. For example, it's more difficult to build a highway in the mountains than on the plains. Inside the *Best Transportation* folder are five sites for you to evaluate.

3. Which of the placemarked sites in the *Best Transportation* folder has the best existing transportation infrastructure, as defined by access to different types of transportation within 5 kilometers?

   A. A
   B. B
   C. C
   D. D
   E. E

Finally, natural hazards are another major consideration in terms of location and development. Turn off the *Best Transportation* folder and turn on the *Environmental Risk* folder. Certain parts of the world are prone to higher risks from particular natural hazards. For example, you are not going to find an issue with snow avalanches in Florida. In Switzerland, however, people have had to take numerous precautions to mitigate the risks associated with these hazards, such as building avalanche sheds for cars on mountainous roadways. Look at the locations in the *Environmental Risk* folder and identify the placemark that is <u>not</u> associated with an environmental risk for that location. You might need to do some Internet searches to find out where certain types of hazards are most common.

4. Which hazard is <u>not</u> likely to occur at the placemarked location?

    A. earthquake
    B. sea-level rise
    C. typhoon
    D. tornado
    E. volcano

Exploration 13.3 SHORT ESSAY

1. Revisit the *New Society* folder. Explain the answer you selected by describing the resources visible at your chosen location.

2. Infectious diseases are a major problem for many developing countries, particularly in the tropics. Describe how a country with a high number of persons affected by infectious diseases is handicapped in its efforts to achieve higher levels of development.

Exploration 13.4: US Development

The American Human Development Index is created in a very similar fashion to the UN HDI. A composite of health, education, and income indicators allow states to be ranked from more to less developed.

Exploration 13.4 MULTIPLE CHOICE

1. Turn off the *Environmental Risk* folder and open the *US HDI* folder to determine which of the following statements is most strongly supported. Darker states have lower levels of development while lighter states have higher levels of development.

   A. New York is in the highest tier of human development.
   B. The top ten most developed states are located along coastlines.
   C. The South-central United States generally exhibits lower development scores.
   D. The Northeast generally has lower development scores than the remainder of the country.
   E. Texas is more developed than California.

Turn off the *US HDI* folder. Turn on the *US Gini Index* folder to see a map of income inequality on a state-by-state basis. With this portrayal, darker-colored states are those with greater income equity while lighter states have greater income inequity. Study the map and identify patterns and core areas of more and/or less income equality.

2. Evaluate the statements with regard to the *US Gini Index* map and select the statement that is most strongly supported.

   A. The Inter-Mountain West has several states with higher levels of income equity.
   B. In general, the South has more income equity than the North.
   C. There are more states west of the Mississippi River than east with the highest degrees of inequity.
   D. The gap between rich and poor is greatest in the Midwest.
   E. There is no discernible regionalization in the pattern of income equity in the United States.

Turn off the *US Gini Index* folder and turn on the *Uninsured* folder. The percent of uninsured in a population can be used to reflect the quality of health care in a given region. Do you think the overall health of a population is better or worse when prevention and treatment is affordable? In the United States, the number of uninsured persons ranges from a high of 28 percent in Texas to a low of 6 percent in Massachusetts. The *Uninsured* folder illustrates the pattern across the United States. Darker colors represent states with more uninsured persons and lighter colors represent states with fewer uninsured persons.

3. Identify the region that has the highest percentage of uninsured person as illustrated by the *Uninsured* folder.

   A. Midwest
   B. Southwest
   C. Northeast
   D. Southeast
   E. Northwest

The United States also faces the same hazard risks that others around the world encounter. There is great diversity across the country in terms of likely hazards. Deactivate the *Uninsured* folder and open the *US Hazards* folder to see a sample of hazards and where they are likely to occur.

4. Which of the placemarked locations in the *US Hazards* folder is least likely to face the hazards associated with its placemark?

   A. A
   B. B
   C. C
   D. D
   E. E

Exploration 13.4 SHORT ESSAY

1. Why are some environmental hazards generally less devastating to a country such as the United States than less developed countries? Select a hazard and elaborate.

   _____

   _____

   _____

   _____

   _____

   _____

2. Go to the *US HDI* folder and click the link to the American Human Development Project Web site to assess development characteristics of your location. When the Web site opens, click the Start Exploring button. At the top of the new screen that appears, go to the "Find by zip code" search box and type in your zip code to view information about your congressional district. Compare these results with the values for your state and provide a brief explanation.

_____

_____

_____

_____

_____

_____

_____

Name: _____

Date: _____

Chapter 14:
Agriculture

Without agriculture, our large, urban, and complex societies would not be possible and most people would fail to survive. Agriculture creates the surplus of food that is directly responsible for our highly specialized economies. Agriculture employs more people around the world than any other economic sector, uses more land than any other human activity, and provides us with some of our most distinctive cultural variation. This exploration will examine the hearth areas of agriculture, the differences between agriculture in more and less developed countries, and conclude with some elements that address the future of agriculture.

Download EncounterHG_ch14_Agriculture.kmz from **www.mygeoscienceplace.com** *and open in Google Earth™.*

Exploration 14.1: Origins of Agriculture

Agricultural hearths represent important regions where the domestication of plants and animals began at least 10,000 years ago. The *Hearths* folder identifies several of these cultural regions. Key innovations occurred at each of the locations that laid the groundwork for the remarkably productive agricultural systems of our modern world. Use your textbook to help you understand the unique characteristics of each agricultural hearth.

Exploration 14.1 MULTIPLE CHOICE

1. Which is not considered a global hearth?

   A. A
   B. B
   C. C
   D. D
   E. E

2. Select the hearth region that is home to the domestication of the following suite of crops: sorghum, yams, coffee.

   A. A
   B. B
   C. C
   D. D
   E. E

3. Which area was the region of the most productive animals for agriculture?

    A. A
    B. B
    C. C
    D. D
    E. E

Early agriculturalists in North America worked with a diverse range of resources to establish farming. One intriguing location is Sunset Crater. Go to the *Sunset Crater* placemark to survey the scene. Sunset Crater is a National Monument today. You can learn more about the site from the National Park Service or by utilizing search terms such as "Sunset Crater agriculture" on the Web.

4. Using your visual assessment of the scene and any supplemental information you locate from outside resources, determine which of the following statements is most strongly supported.

    A. Sunset Crater is a meteorite impact zone that created mineral-rich soils for farming.
    B. Sunset Crater was home to the first wheat growers in the Americas.
    C. This was an early site of agriculture because of the abundance of moisture and streams in the area.
    D. Early agriculturalists used the cinders of Sunset Crater as mulch to help establish young plants.
    E. Sunset Crater was an early agricultural center because it is located on the US-Mexico border.

Exploration 14.1 SHORT ESSAY

1. The low levels of agricultural productivity in Mesopotamia and the Fertile Crescent today indicate that this is not the most important agricultural region today. What do you think is the most important agricultural region today? Explain your answer.

    _____
    _____
    _____
    _____
    _____
    _____

2. What is an example of an early agricultural practice or technology in or near the location where you live? Is there any reflection of/connection to the historical agricultural activity you cited when you view your area from Google Earth™ today?

    _____
    _____
    _____
    _____
    _____
    _____

Exploration 14.2: Traditional Agriculture

The range of agricultural techniques and methods continues to diffuse around the world in the twenty-first century. Thus it is becoming more difficult to differentiate between more traditional and more modern types of agricultural systems. For example, large, center-pivot irrigation systems are being increasingly used in developing countries. Nonetheless, there are still striking differences to be found around the world. Several agricultural systems are disproportionately associated with developing countries.

Exploration 14.2 MULTIPLE CHOICE

1. Go to the *Ag Scene 1* placemark and identify the agro-ecological system most likely in place at this location.

   A. wet rice farming
   B. shifting cultivation
   C. pastoral nomadism
   D. dairy
   E. commercial gardening

2. Go to the *Ag Scene 2* placemark and identify the agro-ecological system most likely in place at this location.

   A. wet rice farming
   B. shifting cultivation
   C. pastoral nomadism
   D. dairy
   E. commercial gardening

Illicit drug crops are disproportionately grown on large scales in the developing countries of the world. Problem crops can include marijuana, coca (for cocaine), and poppies (for heroin). View the placemarked locations in the *Drug Crops* folder. Think about considerations such as rural locations and the ideal climates/growing conditions for the aforementioned illicit drug crops.

3. Identify the location in the *Drug Crops* folder that is most likely the location of crop production in the developing world to be used for illicit drugs.

   A. A
   B. B
   C. C
   D. D
   E. E

In 2008, Nigeria led the world in the production of several crops. The other top-producing countries (by value of the crop) have been placemarked in the *Mystery Crop* folder. You will likely need to visit the Food and Agricultural Organization of the United Nations. The link associated with the *Mystery Crop* folder will take you to the site where a wide variety of statistics related to agriculture are available.

4. What is the *Mystery Crop*?

   A. apples
   B. cassava
   C. millett
   D. sorghum
   E. yams

Exploration 14.2 SHORT ESSAY

1. Research why the *Mystery Crop* is important to Nigeria. What environmental trait(s) does Nigeria share with the other leading producers of this crop?

   _____

   _____

   _____

   _____

   _____

   _____

   _____

2. FAOSTAT provides time-series and cross-sectional data related to food and agriculture for nearly 200 countries. Return to the FAOSTAT Web page and examine the statistics regarding the top producers for various commodities. Identify a commodity where you can see a direct cultural relationship between the level of production (or lack thereof). For example, Germany leads in production of hops. Hops are an essential ingredient for beer production. Beer is a central component of the German food tradition.

   _____

   _____

   _____

   _____

   _____

   _____

   _____

Exploration 14.3: Modern Agriculture

The United States is a good case study for understanding the characteristics of modernized agriculture. One of these characteristics is specialized monocropping where the same crop is grown in the same field year after year. This practice has the benefits of being efficient as specialized equipment and fertilizers allow for the benefit of producing at a very large scale. Monocropping, along with climatological and pedological (soil) influences, contributes to vast areas of the United States specializing in one particular crop.

Open the *Modern Agriculture* folder. The *Crop One* and *Crop Two* image overlays illustrate the distribution of two crops that are commonly monocropped in the United States. Visit the US Department of Agriculture Web site associated with the *Crop One* placemark and use your textbook and any outside resources to determine what crops are displayed.

Exploration 14.3 MULTIPLE CHOICE

1. Identify the crop distribution displayed by the *Crop One* image.

    A. corn
    B. lettuce
    C. soybeans
    D. tomatoes
    E. wheat

2. Identify the crop distribution displayed by the *Crop Two* image.

    A. corn
    B. lettuce
    C. soybeans
    D. tomatoes
    E. wheat

Industrialized agriculture provides distinctive cultural landscapes beyond monocropping. Go to the *Industrialized Agriculture 1* placemark to view a representative scene of agriculture in the developed world.

3. Use your textbook and any outside resources to determine what term or phrase best describes the scene.

    A. genetically modified organism
    B. concentrated animal feeding operation
    C. aquaculture
    D. ethanol production
    E. slash and burn

Another major component of industrialized agriculture is the commodities market. Commodities, or tradable goods with an established demand, include a number of agricultural products such as wheat. Go to the *Industrial Agriculture 2* placemark to view an important element of the wheat commodity market. The white cylindrical structures are grain silos. Great amounts of wheat are stored here. At times, wheat will be stored for longer periods to wait for prices to increase. View the scene around the elevators. Zoom out to see the larger spatial context.

4. What visible element of the landscape surrounding the *Industrial Agriculture 2* is most essential for the wheat commodities market?

    A. railroads
    B. navigable river
    C. airport
    D. port
    E. residential housing

Exploration 14.3 SHORT ESSAY

1. The trend toward bigger and bigger farms has been ongoing since the beginning of agricultural domestication. Organic farms are one example that has bucked this trend to some degree. Locally grown foods are also gaining in popularity and often produced by smaller agricultural operations than the mega-farms. Identify evidence of small farms or agricultural plots in Google Earth™ near your location. Describe what you see.

   _____
   _____
   _____
   _____
   _____
   _____
   _____

2. The *Meat Packing* placemark reflects an activity that has become increasingly common on the American Great Plains in recent decades. These facilities process enormous amounts of beef raised through the use of modern intensive agricultural practices. In fact, you can navigate your way just a few miles west of the *Meat Packing* placemark to see another example. As these facilities have increased in number, there has been a dramatic demographic shift in many communities of the region. Complete any necessary research in order to describe the dominant demographic change.

   _____
   _____
   _____
   _____
   _____
   _____
   _____

Exploration 14.4: Future of Agriculture

Aquaculture is one of the ways that people will continue to work toward ever-greater food production. Aquaculture involves the production of freshwater and saltwater fish and crustaceans in controlled environments. Aquaculture has represented one way forward for the modernization of developing countries' agricultural systems. For example, China and Southeast Asian countries such as Thailand and Vietnam are major exporters of farm-raised fish and crustaceans. While economic development and an increased food supply represent general benefits of this rapidly growing industry, there are negative aspects, as well.

Exploration 14.4 MULTIPLE CHOICE

1. Based on your analysis of the *Aquaculture* placemark and any outside research you have completed, identify the response that is <u>not</u> an issue associated with this type of aquaculture.

   A. Farmed fish could escape from pens and breed with their wild companions.
   B. A high amount of waste is generated that must be discharged into the surrounding environment.
   C. Large numbers of fish in such close proximity dictate the use of large amounts of antibiotics.
   D. Where aquaculture is practiced, all other types of organized agriculture are abandoned.
   E. Aquaculture can have a more severe detrimental environmental impact at the local scale than the exploitation of wild fisheries.

As the need for agricultural production increases, more land has been dedicated to this purpose. The world's forests represent common targets for agricultural expansion. For the past several decades there has been much attention paid to the clearing of forest in the Amazon Basin for agricultural purposes. Go to the *New Agriculture* placemark and view this transition underway. Study the area surrounding the placemark and complete some outside research on the root causes of Amazon deforestation. Be sure to utilize the historical imagery feature in your assessment of the landscape.

2. Identify the statement that is supported by using Google Earth™ imagery and your outside research.

   A. The area around the *New Agriculture* placemark is one part of the Amazon not affected by agricultural expansion.
   B. Agricultural development is correlated most strongly with the location of streams.
   C. A principal driver of forest clearance in this region is the creation of pasture for cattle.
   D. Wheat is the primary crop being planted in the region of the Amazon River Basin.
   E. All of the forest parcels within 10 kilometers of the *New Agriculture* placemark have been cleared for agriculture.

Changing global agricultural patterns also reflect the increasing disparity between the most affluent and poorest populations. Specialized agricultural production has targeted wealthy consumers in the developed nations of the world, resulting in the creation of various associations between regions and specific agricultural products in recent decades. One of these emerging, product-specific regions is Southeast Australia. Go to the *Luxury* placemark and survey the cultural landscape. If you are unable to discern what type of product is being produced here, search the Web for information on the agriculture of Southeast Australia.

3. What product is being produced at the *Luxury* placemark?

   A. Wine
   B. Kobe Beef
   C. Cheese
   D. Foie Gras
   E. Caviar

In recent years there have been some interesting trends in consumer produce preferences in the developed world. Organic products have become increasingly popular. Just because food is organic, it does not necessarily equate to the most environmentally friendly choice. For example, many of the fruits and vegetables we may purchase at the grocery store have been shipped great distances from locations where a given crop is in season. Therefore, your salad can have a significant carbon footprint. This is one reason why local foods have become an increasingly popular choice for some consumers. Go to the *First Class Salad* placemark and calculate the air miles that would be calculated by creating a salad of spinach, strawberries, avocado, and mango. Use the measure tool and add the straight-line distances from the following sources to Fargo: Spinach (Imperial County, California), Strawberries (Irapuato, Mexico), Avocado (Cariu, Indonesia), and Mango (Pondicherry, India).

4. What is the approximate collective air mileage of the ingredients of the salad that was assembled in Fargo?

   A. 2,000 kilometers
   B. 8,000 kilometers
   C. 17,000 kilometers
   D. 28,000 kilometers
   E. 61,000 kilometers

Exploration 14.4 SHORT ESSAY

1. A sizable percentage of the corn produced in the United States is sent to ethanol plants to be converted to a corn-based fuel additive. Go to the *Ethanol Plant* placemark to see an example of one of these facilities. Comment on the location of this facility. Why is it here? Then add a comment on whether you support food being utilized for fuel.

   _____
   _____
   _____
   _____
   _____
   _____
   _____

2. Go to the *Unique Agriculture* placemark. You will see a distinctive agricultural landscape. While this area has produced this crop for a very long time, the industry is in a period of noteworthy change. Determine what is being grown here, if the demand in the last decade has been increasing or decreasing, and what the prospects for this type of agriculture are in the near future. The *Photos* layer may be helpful to you to answer this question.

   _____
   _____
   _____
   _____
   _____
   _____
   _____

Name: _____

Date: _____

Chapter 15:
Industry

The diverse activities of industry are shaped by a range of cultural facets. Our preferences for the products we need and desire are impacted by not only our beliefs and values, but also by popular culture. Economies around the world work to meet these demands through the processes of industry. From the extraction of the raw materials to the processing of these materials to the manufacture, distribution, and consumption of goods, the economic machine hums around us without pause. In this exploration we will focus on primary and secondary industries, with an emphasis on the locational factors that help explain the distribution of these activities.

Download EncounterHG_ch15_Industry.kmz from **www.mygeoscienceplace.com** *and open in Google Earth™.*

Exploration 15.1: Primary and Secondary Industries

Primary industries extract natural resources. Because natural resources are spread unevenly around the globe, there are clusters of extraction activity for different types of resources. Having access to a wide range of such natural resources is tremendously beneficial to a national economy. One of these important resources is timber. Timber is valuable not only as a building material, but also as a fuel source. While timber is no longer relied upon to fuel the furnaces of the developed world, it remains an important primary industry. Open the *Primary and Secondary Industry* folder, and go to the *Washington State* placemark to view a timber-based industrial landscape. Use the Historical Imagery tool to assess the ways that the landscape has or has not changed in recent years.

Exploration 15.1 MULTIPLE CHOICE

1. What statement can you verify by studying the evolution of the landscape in the scene around the *Washington State* placemark?

   A. Timber has not been harvested here in since 1998.
   B. Timber is harvested from a single forest road that has been constructed through the forest.
   C. Based on re-growth, timber has been harvested at two different times in the past.
   D. Once a plot has been harvested, there are no signs that timber re-grows.
   E. A wide variation in spatial and temporal activity is associated with this primary industry.

Turn off the historical imagery. In the early days of the Industrial Revolution, several industries were impacted dramatically as changes in techniques and technology led to more efficient methods of production. Evolving industries included iron, coal, transportation, textiles, chemicals, and food processing. Go to the *Garzweiler, Germany* placemark to view a distinctive industrial landscape related to one of these industries. This is an area that would have been important for the production of a key component for industrialization. Of course, the scale of the production is much larger today.

2. Assess the landscape around the *Garzweiler, Germany* placemark and determine which one of the following industries is represented here.

    A. chemicals
    B. textiles
    C. coal
    D. food processing
    E. iron

Secondary industry represents the next step in the industrial process. In secondary industry, the raw goods are processed and manufactured into finished goods. This can include complex products such as aircraft or refined natural resources such as gasoline. Many of these operations utilize a wide array of components that are brought to a central location where the final steps in the manufacturing process take place. One such example is found at the *Secondary Industry 1* placemark. Study the facility that has been placemarked. There are numerous clues visible in Google Earth™ as to what is being manufactured here.

3. What is the secondary industry associated with the *Secondary Industry 1* placemark?

    A. aircraft
    B. automobiles
    C. televisions
    D. petrochemicals
    E. computer chips

A very different industrial landscape is evident around the *Secondary Industry 2* placemark. Go to the *Secondary Industry 2* placemark and view a location that is one of the largest of its kind in the world. Be certain to zoom out to the wider landscape to see some important clues as to the nature of this site.

4. What is the secondary industry associated with the *Secondary Industry 2* placemark?

    A. bottled water plant
    B. oil refinery
    C. brewery
    D. textile manufacturing
    E. appliance manufacturing

Exploration 15.1 SHORT ESSAY

1. Use Google Earth™ to find an example of primary and secondary industry near you. Report the coordinates in latitude and longitude and provide a brief description of the industries that you have located.

_____

_____

_____

_____

_____

_____

_____

2. Go to the *Alaska* placemark. Assess the scene. What is an example of industry that might be occurring at this location? Don't worry about finding the "correct" answer; just make an educated guess on the primary and/or secondary industry that could be based at this location. Be sure to explain your answer.

_____

_____

_____

_____

_____

_____

_____

Exploration 15.2: Spatial Evolution of Industry

The Industrial Revolution refers to the changes that occurred as machines started to do the work of people, thus increasing human productivity exponentially. The developments of key technologies such as the steam engine diffused to other parts of the world over time. Begin by utilizing your text to determine the origin of the Industrial Revolution and the subsequent spatial diffusion. Open the *Spatial Evolution of Industry* folder and then open and turn on the *Industrial Diffusion* folder.

Exploration 15.2 MULTIPLE CHOICE

1. Track the diffusion of the Industrial Revolution from its point of origin to other world regions by placing the placemarks in the *Industrial Diffusion* folder in a logical order.
   A. 1,2,3,4
   B. 2,3,1,4
   C. 1,3,2,4
   D. 3,2,4,1
   E. 2,1,3,4

As the Industrial Revolution took root in the United States, locations like Lowell, Massachusetts, became hubs of industrial activity. This was a planned textile manufacturing community established in the 1820s. By the middle of the nineteenth century, Lowell was one of the largest and most important industrial complexes in the United States. Go to the *Lowell, Massachusetts* placemark and turn on the *Photos* layer. View the photos of the area and study the Google Earth™ imagery.

2. Identify the statement that is most strongly supported regarding the area around the Lowell, Massachusetts placemark.

    A. The area is still a major growth center of industry.

    B. The old mills have all been demolished to accommodate housing and retail establishments.

    C. Some of the mills and old buildings have been preserved as they are an important part of Lowell's history.

    D. Lowell was bound to eventually fail because of it lack of access to transport by water.

    E. It appears the canal system in the city is a recent addition in the last several decades.

Textiles are produced around the world in a variety of settings today. For a modern sample of textile production, go to the *Textiles Today* placemark. Needless to say, the scene is quite different than what occurred in Lowell as part of the Industrial Revolution in the nineteenth century. Survey the scene, being sure to utilize the historical imagery capabilities of Google Earth™. Turn off the historical imagery when you have completed your analysis.

3. Based on your analysis of the scene around the *Textiles Today* placemark, what statement is not supported?

    A. This has been a textile center since at least 1995.

    B. The textile industry has expanded here in recent years.

    C. New infrastructure has been built in the area during the last several years.

    D. Textile factories are constructed on plots of land that were used for agriculture.

    E. Textile manufacturing is occurring.

Turn on the *Industry Share of GDP (%)* folder and double-click the folder. This extruded set of polygons illustrates the percentage contribution of manufacturing for each country. Examine the regional patterns and think about reasons why a country displays a particularly low or high value with this dataset. Perhaps the country has an economy that is dominated by the economic sectors of agriculture or services? Turn off the *Industry Share of GDP (%)* folder when you have completed your analysis.

4. Identify the statement that is supported by the *Industry Share of GDP (%)* folder.

    A. China is the world's largest secondary manufacturer.

    B. Africa displays the greatest regional variation.

    C. Europe's economies are the most heavily dominated by the industry sector.

    D. More of the US economy is made up of industry than that of Canada or Mexico.

    E. Japan has the most industry-dependent economy in East Asia.

Exploration 15.2 SHORT ESSAY

1. Compare the industrial landscapes of the *Lowell, Massachusetts* and *Textiles Today* placemarks. Project what these landscapes will look like 50 years from now. You could consider factors such as building materials as well as general economic trends that could affect these locations.

_____

_____

_____

_____

_____

_____

_____

2. Return to the *Industry Share of GDP (%)* folder. Pick one distinctive pattern that you see and explain the root causes. For example, you could explain why many of the most affluent countries in the world have industry sectors in the vicinity of 25 percent. Turn off the *Industry Share of GDP (%)* folder when you have completed your analysis.

_____

_____

_____

_____

_____

_____

_____

Exploration 15.3: Situation Factors

Situation factors refer to the geographic context surrounding a site of industrial production. We will look at site factors in the next section. Here we need to think about issues such as the transportation of materials to and from a factory. Beyond the distance to the factory or processing center, the manufacturer must also consider the distance to the prospective customer. Additionally, it's not as simple as calculating distance in most cases. There can be many factors that make the situation more difficult. For example, open the *Situation Factors* folder and go to the *Difficult Situation* placemark. This is a diamond mine in Canada. You may find it useful to use the *Photos* and *Roads* layers for this question.

Exploration 15.3 MULTIPLE CHOICE

1. Survey the situation and identify the statement that is most strongly supported.

   A. There are a number of roads that connect to Canada's highway system.
   B. Ore is likely shipped to a processing plant by sea.
   C. The river is the key way materials are moved in and out of this remote site.
   D. An ice road, constructed each winter, is the most likely way heavy materials and equipment can move to and from the site.
   E. Trains likely move the diamond ore to a remote processing plant.

An entirely different perspective on situation factors comes from the consideration of agglomeration effects. Agglomeration, or clustering, occurs when a number of businesses locate near one another to take advantage of the support systems that arise to support similar industries. This can snowball as workers with expertise in the field move between companies and other businesses locate in the area to take advantage of the clustering of potential customers. Go to the *Agglomeration* placemark.

2. Use the *Find Businesses* tab in the *Search* pane of Google Earth™ to determine which of the following industries have agglomerated at this location. Hint: Don't be fooled by retail establishments.

   A. logging
   B. computer
   C. textile
   D. furniture
   E. mining

Head down the coast by clicking on the *Long Beach, California* placemark. This is one of the largest ports in the world. An unfathomable amount of manufactured goods are shipped into the United States here and then distributed by railroads and highways. Likewise, many manufactured goods leave the United States from this location. Zoom in and examine the site. You will see numerous rail lines and highways that converge here. There is one element, more than any other, that helps this movement of goods to proceed as efficiently as possible.

3. Select the visible element that contributes most strongly to the efficient transfer of goods at the *Long Beach, California* port and others like it around the world.

   A. democracy
   B. numerous cranes on one dock
   C. standard gauge railroad
   D. standardized cargo containers
   E. deepwater port

4. Access to transportation is an important part of the "big picture" of manufacturing and should be considered when any company is assessing the geographic situation of a potential point of production. Imagine that you have a very specialized manufacturing company. Overnight delivery is an essential service for your products. Of the locations contained in the *Overnight Delivery* folder, what would be the best location for your secondary industry?

   A. A
   B. B
   C. C
   D. D
   E. E

Exploration 15.3 SHORT ESSAY

1. Much of the historic American Manufacturing Belt now looks like the scenes contained in the *Rust Belt* folder. What factors explain this?

_____
_____
_____
_____
_____
_____
_____

2. The ships leaving the port in Long Beach have very different cargo than the ships leaving the ports of East Asia headed toward the United States (see the *Hong Kong* placemark for an example port). What do you think are typical items to be found in the shipping containers of the ships going to China from the United States and vice versa?

_____
_____
_____
_____
_____
_____
_____

Exploration 15.4: Site Factors

In comparison to situation factors, site factors are a more direct reflection of the immediate surroundings. Many industrial sites could not be successful without the unique combination of characteristics of a given location. Facets of land, labor, and capital are usually considered when evaluating sites. Open the *Site Factors* folder and go to the *Saltworks* placemark to see a site of early industrial development in France. Assess the site in Google Earth™ and use the associated link to read the information at the UNESCO World Heritage Web site.

1. Identify the single most important historical site factor for salt production at Salins-les-Bains, France.

   A. easy access to a local aquifer
   B. surface salt deposits
   C. proximity to Bracon
   D. the Chaux Forest
   E. highway connectivity for access to international markets

Go to the *Aluminum Smelter* placemark. The large building in the scene is an aluminum smelter. A key site factor for aluminum smelting is present in the view. Research the term "aluminum smelting" on the Web.

2. What is the key site factor for aluminum smelting that is present in the Google Earth™ imagery?

    A. nearby source of electricity
    B. a large educated workforce
    C. location on a bluff
    D. source of wood to generate high heat
    E. navigable stream to move raw materials to smelting plant

In today's global marketplace, situation and site factors work together to create manufacturing that is much more mobile than in the past. Companies relocate often in search of the combination of the site and situation factors that will maximize profits. Go to the *Secondary Manufacturing Site* placemark to see a typical industrial scene from Southeast Asia. Assume that this is a North American–based corporation that has determined this is the best place to manufacture its product.

3. Which of the following site factors is likely most responsible for a secondary manufacturing industry that produces goods for the North American market locating at the *Secondary Manufacturing Site*?

    A. high government taxes
    B. low labor costs
    C. close proximity to market
    D. strong unions / worker protection laws
    E. restrictive environmental regulations

Site and situation factors do not operate independently of one another, but rather in concert. Go to the *Industry Center* folder to see an example of an industrial phenomenon that illustrates this nicely. View the placemarked features inside the folder.

4. Select the industry-related term that best describes these placemarked features.

    A. maquiladoras
    B. cottage industry
    C. break-of-bulk point
    D. guild industry
    E. deindustrialization

Exploration 15.4 SHORT ESSAY

1. Go to the *NASA* placemark. What site factors do you think may have contributed to NASA selecting this location as their primary location (Kennedy Space Center) to send objects into orbit around the Earth?

    _____
    _____
    _____
    _____
    _____
    _____
    _____

2. Go to the *Change* placemark. Evaluate the scene and provide a brief synopsis of what is happening here. Is this a location where primary or secondary industry is more prominent? Is this an area that is growing or in decline? Hint: Pay attention to the elevation readings in Google Earth™ around the placemark and also take a look at the historical imagery.

_____

_____

_____

_____

_____

_____

_____

Name: _____

Date: _____

Chapter 16: Services

A service is an activity that caters to the needs and desires of humans. Services are the core component of industrial activity in the developed world and a growing component in the less developed world. Services are disproportionately clustered in settlements around the globe. The explanation of why and how services cluster into settlements is an example of the study of economic geography. This exploration examines some of these patterns while providing examples of the different types of services.

Download EncounterHG_ch16_Services.kmz from **www.mygeoscienceplace.com** *and open in Google Earth™.*

Exploration 16.1: Service Sector

The three primary components of the service sector are consumer services, business services, and public services. Examples of consumer services include education, health, leisure, and retail and wholesale sales. Business services support other businesses and include professional, financial, and transportation services. Public services are government-related services such as police and fire protection. Read your textbook to develop a more nuanced understanding of the different types of services provided within the service sector. Open the *Service Sector* folder and go to the *Consumer Service* placemark.

Exploration 16.1 MULTIPLE CHOICE

1. Identify the dominant type of consumer service visible in the *Consumer Service* street view.

    A. professional
    B. education
    C. health
    D. leisure
    E. wholesale

Now fly to the *Business Service* placemark to view a different component of the service sector.

2. Identify the dominant type of business service visible in the *Business Service* placemark.

    A. financial
    B. transportation
    C. retail
    D. information
    E. professional

Turn on and open the *National Leaders* folder. Several of the states that have been flagged in the *National Leaders* folder lead the country in number of people per 1,000 that work in select service industries. These industries include the following: fashion designers (0.82 per 1,000 in this state, 0.12 per 1,000 on in the United States), gaming dealers (18.99, 0.67), and meeting and convention planners (3.03, 0.39). Draw on your personal knowledge and any outside resources to identify the states associated with each industry. If you need help, you can find this data and much more at the United States Bureau of Labor Statistics. When you have completed the following question, turn off and close the *National Leaders* folder.

3. Identify the states and/or District of Columbia in the *National Leaders* folder associated with the highest proportions of workers in the following service industries: fashion designers, gaming dealers, meeting and convention planners.

    A. 2,1,4
    B. 3,2,1
    C. 4,1,5
    D. 1,2,3
    E. 5,4,1

Like so many other cultural phenomenon, regional patterns are evident when surveying the percentage of workers in a country who work in the service sector of the economy. Turn on the *% Labor Force in Services* folder and examine the global patterns and outlier countries. When you complete the following question, turn off the *% Labor Force in Services* folder.

4. Identify the statement that is most strongly supported by the *% Labor Force in Services* folder.

    A. In the more affluent countries, a smaller percentage of the labor force work in services.
    B. In general, the countries with the lowest percentages of laborers in service are found in Asia.
    C. Western European countries have higher percentages of laborers in the service sector than Eastern European countries.
    D. Libya has the highest service percentage of laborers on the African continent.
    E. The United States' workforce is the most heavily weighted toward service of any country.

Exploration 16.1 SHORT ESSAY

1. Find your current place of residence on Google Earth™ and identify the three closest service entities. List each service and identify the type of service (e.g., consumer (retail)).

   _____
   _____
   _____
   _____
   _____
   _____
   _____

2. Use Google Earth™ to examine your community. Identify and describe two service facilities that have unique or distinctive footprints in some way.

   _____
   _____
   _____
   _____
   _____
   _____
   _____

Exploration 16.2: History of Services

Services were central components of the first urban settlements to arise around the globe. One service in particular has been cited as a focus of many early settlements. This service was often represented in early settlements by a centralized building of considerable height and size. Open the *History of Services* folder and then open and turn on the *First Service* folder. Examine the five modern street scenes contained in the *First Service* folder. When you complete the following question, turn off the *First Service* folder.

Exploration 16.2 MULTIPLE CHOICE

1. Which scene from the *First Service* folder is analogous to what many social scientists believe was the first service that brought people together in early settlements?

   A. A
   B. B
   C. C
   D. D
   E. E

As early settlements evolved, different types of consumer, public, and business services began to arise in settlements. Turn on the *3D Buildings* layer and go to the *Early Service* placemark. Note the 3D rectilinear feature displayed.

2. What essential service type was most directly served by a layout (focusing on the 3D rectilinear feature) like the one presented at the *Early Service* placemark?

    A. public
    B. financial
    C. warehousing
    D. hospitality
    E. information

The most critical early business service was focused on an essential element of daily life needed for survival. Open and turn on *Early Business Services* folder and view the five modern scenes contained therein. One of these scenes provides a clear modern example of this essential business service. You may find it helpful to turn on the *3D Buildings* layer. Turn off the *Early Business Services* folder when you complete the following question.

3. Select the modern analog from the *Early Business Services* folder that is the best example of an early business service.

    A. A
    B. B
    C. C
    D. D
    E. E

Medieval urban settlements in Europe often had a common service-based urban form that consisted of a central square or plaza with churches, palaces, and other important buildings on the periphery. Clustered around this central space, medieval cities were densely constructed with narrow, winding streets contained within an exterior wall. Turn on and open the *Medieval Settlement* folder and view the placemarked settlements. When you have completed the following question, turn off the *Medieval Settlement* folder and the *3D Buildings* layer.

4. What placemark from the *Medieval Settlement* folder best represents a Medieval European urban settlement?

    A. A
    B. B
    C. C
    D. D
    E. E

Exploration 16.2 SHORT ESSAY

1. Think of the earliest services that were provided in the history of your community. Use Google Earth™ to locate one of these sites. What is the service? Comment on the geographic influences that could have influenced the specific selection of this site.

_____
_____
_____
_____
_____
_____
_____

2. Turn on the *3D Buildings* layer and fly to the *Early Urban Settlement* placemark. What is this location? In what modern country is it located? What is its significance in terms of services and urban history? Turn off the *3D Buildings* layer when you are done.

_____
_____
_____
_____
_____
_____
_____

Exploration 16.3: Service Patterns and Forms

Services are frequently clustered in settlements. Rural settlements have fewer services available than urban settlements and are often focused on agricultural production. The need to share common resources can encourage distinctive patterns such as circular or linear clustering. Open the *Service Patterns and Forms* folder and then open and turn on the *Settlement* folder to view a variety of settlement patterns. Turn off the *Settlement* folder when you complete the following two questions.

Exploration 16.3 MULTIPLE CHOICE

1. Identify the best example of a linear rural settlement from the *Settlement* folder.

   A. A
   B. B
   C. C
   D. D
   E. E

2. Identify the best example of a dispersed rural settlement from the *Settlement* folder.

   A. A
   B. B
   C. C
   D. D
   E. E

The area that surrounds a location where services are obtained is known as a hinterland or market area. Most people will obtain services from local establishments rather than traveling to another market area. For example, gasoline is available at the nodes located roughly in the center of each market area. Open and turn on the *Market Areas* folder to view five hypothetical market areas delineated by polygons. To better see the nodes, you can search for "gas" in the Find Businesses tab of the Search pane. Turn off the *Market Areas* folder when you complete the following question.

3. Select the market area from the *Market Areas* folder that is the most appropriate delineation of a market area for gasoline.

   A. A
   B. B
   C. C
   D. D
   E. E

Turn on and go to the *Range* polygon. The range is the maximum distance people are willing to travel for a service. This hypothetical range is based on a service offered in the Dallas–Fort Worth metropolitan area. Turn off the *Range* polygon when you complete the following question.

4. What service would most justify the market area associated with the *Range* polygon?

   A. Bank
   B. State supreme court
   C. Home electronics store
   D. Lamborghini automobile dealership
   E. NBA professional basketball team

Exploration 16.3 SHORT ESSAY

1. Think of one of the highest order goods available in your community. Locate it in Google Earth™. Does it have a distinctive spatial pattern? Why or why not?

   _____
   _____
   _____
   _____
   _____
   _____

2. The threshold is the minimum number of people needed to support a particular service. Open the *Threshold* folder and examine the placemarked locations. Based on the characteristics of the locations and/or size of the communities, explain the reasons you think the thresholds for the services labeled with the placemarks would or would not be met here.

_____
_____
_____
_____
_____
_____
_____

Exploration 16.4: Location and Services

Open the *Location and Services* folder and then open and turn on the *Optimal Location* folder. Providers of services must decide where to locate their service and if the threshold and range are sufficient to justify their services. Some services are more acutely dependent on their specific location to achieve success than others. For example, imagine a restaurant designed to capitalize on walk-up business. It needs a location that will have the highest levels of foot traffic. Look at the potential locations of this hypothetical restaurant in the *Optimal Location* folder.

Exploration 16.4 MULTIPLE CHOICE

1. Identify the best location in the *Optimal Location* folder for a new restaurant dependent on walk-up customers.

   A. A
   B. B
   C. C
   D. D
   E. E

2. Cities often specialize in different types of services. For example, some cities specialize in manufacturing, while others may be banking or information service centers. Open and turn on the *Montana* folder and view the selected cities to determine which city is the focal point of public services for the state. Turn off the *Montana* folder when you complete the following question.

   A. A
   B. B
   C. C
   D. D
   E. E

Go to the *Service Site* placemark and assess the site. Be sure to utilize the historical imagery tool in Google Earth™ to assist your study.

3. Select the service-related term that best applies to the *Service Site*.

   A. back office
   B. periodic market
   C. central place
   D. hinterland
   E. offshore center

World cities are centers of national and international power. Political and corporate power is clustered in these locations. As a result, the highest order business services conglomerate there. Dominant world cities are home the top stock exchanges and have very large concentrations of financial services. Open and turn on the *World Cities* folder. Turn off the *World Cities* folder when you complete the following question.

4. Identify the placemarked city in the *World Cities* folder that is the best example of a dominant world city.

   A. A
   B. B
   C. C
   D. D
   E. E

Exploration 16.4 SHORT ESSAY

1. Go to the *Rural Service* placemark. What type of service activity is occurring at this location? What do you think is spread out on the ground? If you had access to an image of the location on the next day, what do you think it would look like?

   _____
   _____
   _____
   _____
   _____
   _____
   _____

2. Back office business functions refer to business services such as clerical activities that are increasingly being outsourced to locations in the global periphery. The primary advantage gained by moving these activities overseas is the dramatic decrease in labor costs. You may have experienced this phenomenon when you have called a company for customer service and find yourself speaking to someone in South Asia, for example. Open the *Back Office* folder. Which location is best for a back office operation that will require a high degree of contact with customers based in the United States? Explain the pros and cons of each site.

   _____
   _____
   _____
   _____
   _____
   _____
   _____

Name: _____

Date: _____

Chapter 17:
Urban Geography

What is a city? Governments around the world don't agree on what constitutes a city. In fact, most of us probably have differing notions of what constitutes a city. One thing that is easier to agree upon is the idea that the city is a dynamic entity. Cities change through time and space. The following explorations are designed to give you a sense of the city by examining the clustering that occurs downtown, the organization of people within the city, and the unique settings of both inner cities and the suburbs.

Download EncounterHG_ch17_Urban_Geography.kmz from **www.mygeoscienceplace.com** *and open in Google Earth™.*

Exploration 17.1: Downtown Clustering

The central business district (CBD) is a dense cluster of businesses and offices that is often centrally located within a city. The CBD is usually well-served by transportation corridors and vertical development (e.g., skyscrapers) is common as a result of more valuable land. Open the Urban *Geography* folder, then the *Downtown Clustering* folder, the open and activate the *Kansas City, MO CBD* folder. It contains five polygons (*A* through *E*) that you can identify by turning the polygons on or off. Evaluate the urban environment of the Kansas City metropolitan area to determine which one of the polygons best represents the central business district.

Exploration 17.1 MULTIPLE CHOICE

1. Which polygon best defines the CBD of Kansas City, Missouri?

   A. A
   B. B
   C. C
   D. D
   E. E

The *Minneapolis, Minnesota CBD* folder illustrates the high density characteristics of typical CBD's. However, Minneapolis has a plethora of unique structures found throughout the CBD. These structures would not make economic sense in a lower density area and would not be as beneficial in other similar cities. Zoom in to the placemarks in the *Minneapolis, Minnesota CBD* to view these structures.

2. The structures placemarked by the white pushpins in the *Minneapolis, Minnesota CBD* folder are a result of unique location characteristics related to

   A. security.
   B. trade.
   C. ethnicity.
   D. topography.
   E. climate.

The core of the city in other parts of the world can look very different than what is found in the United States. For example, the built environment in the heart of Islamic cities reflects the emphasis on personal privacy rather than outward appearance. This is one way that gender roles are clearly defined in many Islamic cities.

3. Utilize the selection of placemarks and 360° City views contained in the *Islamic Cities* folder and any outside resources to assess which of the following statements regarding Islamic cities is least accurate.

   A. Doors are sometimes staggered to not face each other across small streets.
   B. Street-level windows are usually smaller than higher windows.
   C. Larger homes are sometimes built around courtyards to create a private space.
   D. Individuality is valued in the outward design of residences in Islamic cities.
   E. Cul-de-sacs are sometimes used to reduce the number of persons passing by homes.

Historically, cities have been located at places that took advantages of site characteristics and situation characteristics. The site refers to the local geographic setting, such as a protected bay. The situation refers to the regional setting or the location of a place relative to other places. One common type of local geographic setting is a site that maximizes defense of the city. Examine the labeled city sites in the *Defensive Site?* folder.

4. Which site in the *Defensive Site?* folder offers the best natural defenses?

   A. A
   B. B
   C. C
   D. D
   E. E

Exploration 17.1 SHORT ESSAY

1. Compare the network of stations for the *London Underground (The Tube)* and the *Toronto Subway* by clicking on each folder. Which of the systems seems to be more focused on the CBD? Do you think one system is older than the other? Why? Note: The London map portrays station locations only.

   _____
   _____
   _____
   _____
   _____
   _____
   _____

2. Open the *Skylines* folder and turn on the *3D Buildings* layer. Compare the urban scenes from *Manhattan* and *Washington D.C.* Then examine the *Paris* placemark. Is Manhattan or Washington D.C. more like Paris? What is at the root of the similarities or dissimilarities? Turn off the *3D Buildings* layer when you have completed your analysis.

_____

_____

_____

_____

_____

_____

Exploration 17.2: People within Urban Areas

Models of urban structure can help make sense of the complex urban environment by categorizing different parts of the city based on who lives there or what takes place there.

Exploration 17.2 MULTIPLE CHOICE

1. Evaluate the polygons in the *Regions within the City* folder and identify which region is associated with each polygon. You may find it useful to zoom in to the Street View imagery at several locations within each polygon.

   A. A. 1. CBD 2. High-class residential 3. Low-class residential 4. Transportation and industry 5. Middle-class residential
   B. B. 1. Middle-class residential 2. High-class residential 3. Low-class residential 4. Transportation and industry 5. CBD
   C. C. 1. CBD 2. Transportation and industry 3. Low-class residential 4. High-class residential 5. Middle-class residential
   D. D. 1. Middle-class residential 2. CBD 3. Low-class residential 4. High-class residential 5. Transportation and industry
   E. E. 1. Middle-class residential 2. Transportation and industry 3. Low-class residential 4. High-class residential 5. CBD

The Laws of the Indies guided the development of Spanish colonial cities to conform to a strict set of standardized plans. Gridiron street plans (ordered arrangement of streets in a grid pattern) with a central church and plaza were responsible for creating homogenous urban environments compared to the older indigenous layouts.

2. Open the *Laws of the Indies* folder and examine the placemarked locations. Identify which location appears to have been least influenced by the Laws of the Indies.

   A. Jinotega, Nicaragua
   B. Los Andes, Chile
   C. Mesilla, New Mexico
   D. Taos Pueblo, New Mexico
   E. Turrialba, Costa Rica

A central focus of many urban settings is public space. From the Greek agora to the Roman forum to the plaza during Spanish colonial times to the county courthouse square in many US cities, public space is accessible by all of the citizens of a city without restriction. Restrictions could include entry fees or owners who decide what is and is not permissible in the space.

Examine the selected locations in and around Orlando, Florida in the *Public Space* folder.

3. Which of the placemarked locations in the *Public Space* folder is the largest public space?

    A. Citrus Bowl
    B. Amway Center
    C. Disney Epcot
    D. Festival Bay Mall
    E. Cherokee Park

Not only do similar socioeconomic groups cluster within cities, but certain types of businesses do as well. This agglomeration, or clustering of similar businesses, provides mutual benefits for businesses. For example, retail stores locate adjacent to one another to take advantage of foot traffic generated by other nearby stores.

4. Go to the agglomeration placemark and determine what type of business/activity has located adjacently at this location.

    A. Movie theatres
    B. Hospitals/healthcare
    C. Warehouse distribution centers
    D. Schools/education
    E. Automobile dealerships

Exploration 17.2 SHORT ESSAY

1. Find your city in Google Earth. What evidence can you identify of patterns that reflect an urban model such as the multiple nuclei model, the sector model, or the concentric zone model?

    _____
    _____
    _____
    _____
    _____
    _____
    _____

2. Examine Fez, Morocco. Describe the patterns you see in the urban landscapes contained within the western and eastern polygons. Would you characterize the respective polygons as European-influenced colonial layouts or traditional Islamic city layouts?

    _____
    _____
    _____
    _____
    _____
    _____
    _____
    _____

Exploration 17.3: Inner City Challenges

Many inner city neighborhoods in the US face tremendous challenges in the 21st century. Turn on the *Street View* layer and then view the sample of neighborhoods contained in the *Inner City Neighborhoods* folder to see some examples of inner city blight.

Exploration 17.3 MULTIPLE CHOICE

1. Which of the following best describes the locations contained in the *Inner City Neighborhoods* folder?

   A. Vacant lots are rare in the highlighted neighborhoods.
   B. There is no evidence of rebuilding or renewal in the highlighted neighborhoods.
   C. There is no evidence of abandoned buildings in the highlighted neighborhoods.
   D. Public infrastructures, such as streets and sidewalks are well maintained.
   E. Trash is a prevalent feature in the highlighted neighborhoods.

Neighborhood Scout (www.neighborhoodscout.com) is a service that integrates geographic, social, economic, and census data in order to provide a snapshot of information for a particular neighborhood. *Neighborhood Scout* uses census tracts as the unit of spatial analysis. Census tracts are statistical subdivisions of a county that usually have between 2,500 and 8,000 inhabitants.

2. Use the *Neighborhood Scout* Web page to identify the statement that is least accurate. (You can use street addresses obtained from Google Earth Street View.)

   A. The educational level of the Central Parkway/Liberty Street neighborhood is lower than the national average.
   B. The income level of the Central Parkway/Liberty Street neighborhood is among the lowest 15% in the US.
   C. Cedar Avenue/55th Street is a mostly black neighborhood.
   D. Mt. Elliott Street/Palmer Avenue is a mostly white neighborhood.
   E. Mt. Elliott Street/Palmer Avenue is a mostly English speaking neighborhood.

There are a number of specialized processes and developments that unfold in urban environments. Utilize your textbook and any outside resources to help you determine appropriate terminology to describe the *Urban Processes 1* and *Urban Processes 2* placemarks.

3. Which of the following terms is most strongly associated with the *Urban Processes 1* placemark?

   A. gentrification
   B. redlining
   C. public housing
   D. filtering
   E. annexation

4. Which of the following terms is most strongly associated with the *Urban Processes 2* placemark?

    A. gentrification
    B. redlining
    C. public housing
    D. filtering
    E. annexation

Exploration 17.3 SHORT ESSAY

Return to the Mt. Elliott Street/Palmer Avenue neighborhood. This is a neighborhood in Detroit, Michigan. Detroit has experienced urban decay as severely as any city in the US. Take a closer look at this neighborhood and surrounding neighborhoods from the bird's eye perspective and the street view.

1. How do you think this neighborhood got to the point it is today? Hypothesize about the transitions that have occurred in the last half-century.

    _____
    _____
    _____
    _____
    _____
    _____
    _____

2. Discuss the potential benefits and drawbacks of gentrification in the form of large-scale festival settings such as those identified within the *Festival Settings* folder.

    _____
    _____
    _____
    _____
    _____
    _____
    _____

Exploration 17.4: Challenges for the Suburbs

Generally, as one travels outward from the core of the city, there is a decline in the population density. Older apartment buildings and row houses eventually give way to large suburban lots. However, the movement of persons within the city that has occurred in the last half century has complicated the picture in some cases.

Exploration 17.4 MULTIPLE CHOICE

1. Open the *Density* folder and order the placemarked residential locations from most dense to least dense. Remember that some buildings, like multi-story apartment structures will contribute to a high density.

    A. 1,4,5,2,3
    B. 3,2,5,4,1
    C. 1,2,3,4,5
    D. D.4,1,2,3,5
    E. 5,4,3,1,2

Over time, the models used to explain cities have changed. The concentric zone model and the sector model may not be as effective as they once were. A common model in use today is the peripheral model proposed by Chauncey Harris.

2. Using Harris' peripheral model as a guide, identify the response that correctly labels residential and commercial features of the Dallas-Fort Worth metroplex.

    A. 1 = office park, 3 = shopping mall, 5 = service center, 4 = industrial district, 2 = airport complex
    B. 2 = airport complex, 3 = central city, 1 = suburban residential area, 4 = office park, 5 = shopping mall
    C. 3 = suburban residential area, 5 = office park, 4 = shopping mall, 2 = airport complex, 1 = industrial district
    D. 1 = service center, 2 = central city, 3 = industrial district, 4 = office park, 5 = airport complex
    E. 5 = industrial district, 1 = airport complex, 2 = central city, 3 = shopping mall, 4 = service center

Examine the schools placemarked in the *School Performance* folder. Make a general assessment of the neighborhood by the bird's eye perspective and street view, and then utilize the *Neighborhood Scout* information to evaluate the schools for likely performance.

3. Which of the schools in the *School Performance* folder are most likely to be highly performing or poorly performing?

    A. highly = 1,2,3; lowly = 4,5
    B. highly = 2,4,5; lowly = 1,3
    C. highly = 1,3,5; lowly = 2,4
    D. highly = 4,5; lowly = 1,2,3
    E. highly = 3,4,5; lowly = 1,2

4. View the placemarks contained in the *Zoning* folder. Combining the information you can gather from a bird's eye perspective along with street views in the immediate vicinity, identify the response that correctly labels the zoning types.

    A. 5 = residential, 4 = industrial, 3 = agricultural, 2 = commercial, 1 = open space
    B. 1 = residential, 2 = industrial, 3 = agricultural, 4 = commercial, 5 = open space
    C. 3 = residential, 2 = industrial, 1 = agricultural, 5 = commercial, 4 = open space
    D. 4 = residential, 3 = industrial, 5 = agricultural, 2 = commercial, 1 = open space
    E. 1 = residential, 2 = industrial, 4 = agricultural, 5 = commercial, 3 = open space

Exploration 17.4 SHORT ESSAY

Edmond, Oklahoma is an affluent and thriving edge city. While it is a prominent suburban node today, this was not always the case. Go to the *Edmond, Oklahoma* placemark and view the urban landscape. Open the *Edmond, Oklahoma 1891* link to view a panoramic map of the area in the period prior to statehood. Manipulate the zoom feature to get a sense of the historic urban environment. Also manipulate the Google Earth view to assess the community at large.

1. Describe some of the differences that are evident between the contemporary and historic urban landscapes of Edmond, Oklahoma.

   _____

   _____

   _____

   _____

   _____

   _____

Use Google Earth to zoom to your city or a city with which you have some familiarity. Ideally, select a metropolitan area with a population of at least 250,000 persons. When the majority of the metropolitan area is in your view, use the *Find Businesses Tab* in the *Search* pane to search for the term "mall."

2. What metropolitan area did you chose? Describe the spatial pattern of the large shopping malls. What element of the urban environment is most strongly associated with mall location? Explain your response.

   _____

   _____

   _____

   _____

   _____

   _____

Name: _____

Date: _____

Chapter 18:
Resource Issues

Our globe has proven to be a rich and diverse stockpile of resources that humans have utilized on their way to higher levels of development. However, these resources are not infinite and have subsequently been degraded or depleted by human activities. In this exploration we examine four categories of resources. Sources of energy, both nonrenewable and renewable, are followed by a look at mineral and biotic resources of earth.

Download EncounterHG_ch18_Resource_Issues.kmz from **www.mygeoscienceplace.com** *and open in Google Earth™.*

Exploration 18.1: Nonrenewable Energy

Nonrenewable energy sources provide the vast majority of the energy needs for the world's people. The big three nonrenewables are coal, oil, and natural gas. The greatest percentage of these resources is consumed by persons living in the most developed countries. Open the *Nonrenewable Energy* folder and turn on the *Oil Production* folder. Evaluate the portrayal of oil producers around the world.

1. What statement is supported most strongly by the *Oil Production* folder?

   A. Every major region of the world has a country in the top ten of oil producing countries.
   B. The production of oil is evenly distributed amongst world regions.
   C. In 2008, Angola was Africa's top-producing country.
   D. Oil is not produced by most European countries.
   E. In 2008, the United States was one of the world's three largest producers of oil.

While mapping the production of oil gives us some idea of what countries are most critical to keeping the global oil-based economic machine churning, mapping the consumption of oil provides a very different distribution. Turn off the *Oil Production* folder and turn on the *Oil Consumption* folder. Study the distribution of global oil consumption. Toggle back and forth between the *Oil Consumption* folder and the *Oil Production* folder to illuminate the largest discrepancies between producers and consumers.

2. Identify the statement that is most strongly supported by the *Oil Consumption* and *Oil Production* folders.

    A. The United States produces about twice as much oil as it consumes.
    B. China is an oil exporter.
    C. Russia is an oil importer.
    D. Norway is an oil exporter.
    E. Brazil has the largest gap between consumption and domestic production.

Natural gas is an increasingly important fossil fuel for the global economy. Natural gas has the benefit of being a much cleaner-burning fuel than oil or coal. There are two folders in the *Nonrenewable Energy* folder related to natural gas: *Natural Gas Imports* and *Natural Gas Exports*. Open these folders and view the global distribution of natural gas importers and exporters.

3. Based on your assessment of the geographical patterns evident in the *Natural Gas Imports* and *Natural Gas Exports* folders, select the most logical statement.

    A. Europe imports natural gas from Russia.
    B. Northern Africa imports natural gas from southern Africa.
    C. Russia imports natural gas from China.
    D. South Korea imports natural gas from Japan.
    E. Saudi Arabia imports natural gas from Iraq.

Turn off the *Natural Gas Imports* and *Natural Gas Exports* folders and open the *Energy Landscapes* folder. This folder illustrates four energy production landscapes. As you can see by these examples, the extraction of nonrenewable energy resources can create impressive changes to the surface of the planet. If you are not familiar with distinctive energy landscapes, turn on the *Photos* layer of Google Earth™ or use the Internet to further research each placemark.

4. Match the energy production landscapes found in the *Energy Landscapes* folder with the type of nonrenewable energy resource associated with each.

    A. 1 = Oil, 2 = Natural Gas, 3 = Coal, 4 = Uranium
    B. 1 = Uranium, 2 = Oil, 3 = Uranium, 4 = Coal
    C. 1 = Natural Gas, 2 = Uranium, 3 = Coal, 4 = Oil
    D. 1 = Coal, 2 = Natural Gas, 3 = Oil, 4 = Uranium
    E. 1 = Natural Gas, 2 = Oil, 3 = Coal, 4 = Uranium

Exploration 18.1 SHORT ESSAY

1. Close the *Energy Landscapes* folder and open the *Mountaintop Removal* folder and turn on the *McRoberts, KY 1983* overlay. This imagery is a product of several organizations' efforts to document the process of mountaintop removal coal mining in the Appalachian Mountains. You are encouraged to explore the wide range of images and information available from the Appalachian Mountaintop Removal layer in Google Earth™ located in the *Global Awareness* folder within the *Primary Database*. For the sake of this question, study the landscape from the 1983 imagery and then turn on the *McRoberts, KY 2004* overlay. Describe the transformation that has occurred and comment on potential impacts to the inhabitants of the region.

    _____
    _____
    _____
    _____
    _____
    _____
    _____

2. Use Google Earth™ to locate the Fukushima Nuclear power plant in Japan that was disabled by the earthquakes and tsunami of March 11, 2011. Then turn on the *Tsunami Forecast Maximum amplitude 3-11-2011* overlay. In what wave height range was the maximum tsunami amplitude that impacted the nuclear plant? Briefly explain what happened to the reactors to cause the eventual partial meltdowns. Comment on the placement of these reactors. Why would they have been located on the coastline? When you have completed the question, be sure to turn off the *Nonrenewable Energy* folder.

    _____
    _____
    _____
    _____
    _____
    _____
    _____

Exploration 18.2: Renewable Energy

In contrast to nonrenewable energy, renewable energy draws on resources that are essentially unlimited in supply. One renewable energy source is hydropower. Hydropower is a major provider of electricity for the world, trailing only coal and natural gas. Energy is generated as water turns turbines. The Three Gorges Dam on the Chang Jiang River in China is the world's largest hydroelectric power station. Open the *Renewable Energy* folder and go to the *Three Gorges* placemark to see the amazing scope of the project. Then move upstream to the *Reservoir* placemark. Utilize the historical imagery capabilities of Google Earth™ to see the post-dam reservoir in comparison to the pre-dam river. You may also want to use the Ruler tool. When you have completed the question, be sure to turn off the Historical Imagery.

Exploration 18.2 MULTIPLE CHOICE

1. Using your assessment of the *Three Gorges* and *Reservoir* placemarks, identify the statement that is most strongly supported.

   A. There is no way for ship traffic to pass the Three Gorges Dam.
   B. The Three Gorges Dam has created a reservoir that stretches at least 250 kilometers upstream.
   C. The Chang Jiang generally flows east to west.
   D. The Three Gorges Dam is more than 4 kilometers in length.
   E. Prior to completion of the reservoir, boats could not navigate the Chang Jiang in the vicinity of the *Reservoir* placemark.

Geothermal energy is energy that is tapped from the interior of Earth. Heat is utilized to create steam for the purposes of turning turbines. The locations where geothermal energy is produced often have a key commonality. Open the *Geothermal* folder and view the three placemarked sites and then consider the location of these sites from a broader geographic perspective.

2. What is the common and relevant physical geography trait for the sites in the *Geothermal* folder?

   A. They are situated along the coastline, where they have immediate access to seawater.
   B. They are located in the most developed countries.
   C. They are in zones with nearby tectonic plate boundaries.
   D. They are in the northern hemisphere.
   E. They are located in warm, tropical climates.

Solar energy provides a variety of opportunities as this energy can be used passively and actively. Active solar involves devices that capture and convert the sun's energy into electricity. Some locations are better situated for making use of this type of solar energy. For example, sunny locations are better than cloudy ones. Turn on the *Annual Solar Radiation* folder to see the best locations in California for active solar collection according to the University of San Diego. If necessary, use the Fly To feature in the Search pane to find cities that are not shown in the initial perspective. When you have completed the question, turn off the *Annual Solar Radiation* folder.

3. What city in California has the highest levels of potential active solar generation?

   A. Fresno
   B. Los Angeles
   C. San Diego
   D. Sacramento
   E. Victorville

Similarly, the potential for the development of wind resources can be evaluated. Turn on the *Hawai'i Wind* overlay to see a map of mean wind speed at a height of 50 meters. This data, provided by the state of Hawai'i, provides a snapshot view of what areas may be acceptable for the development of wind power. Turn off the *Hawai'i Wind* overlay when you complete the question.

4. On which island is the most extensive area with an average wind speed greater than 9.5 meters per second?

    A. Hawai'i
    B. Lana'i
    C. Maui
    D. D.Moloka'i
    E. O'ahu

Exploration 18.2 SHORT ESSAY

1. The boom period of dam building in the United States occurred from the 1920s to 1960s. Numerous large dams were completed in the western United States during the era. One of the most noteworthy is Glen Canyon Dam, which impounds Lake Powell. Go to the *Glen Canyon Dam* placemark to see the site. This dam, like many in the American West, has been controversial for a number of reasons. It is a central figure in the Edward Abbey book *The Monkey Wrench Gang*. Research the dam and the arguments made by its critics. One place to start is the Glen Canyon Institute. What are some of the arguments for the restoration of Glen Canyon and the decommissioning of Glen Canyon Dam?

_____

_____

_____

_____

_____

_____

2. Other renewable energy sources also face criticism. For example, there has been an ongoing battle in the northeastern United States over the last several years over the prospect of a new renewable energy project being constructed. Turn on and go to the *Renewable Controversy* polygon. Complete some research about the controversy on the Internet. What type of renewable energy is targeted for development here? Why has opposition been so strong? What is the status of the project?

_____

_____

_____

_____

_____

_____

Exploration 18.3: Mineral

The earth has an abundance of metallic and nonmetallic mineral resources. While a much higher gross amount of nonmetallic minerals are used by humans, the metallic minerals often are the mineral resources with the highest economic value. Metallic minerals are often crucial components of specialized industrial equipment and manufacturing processes. For example, gold is a precious metal whose price has soared in recent years. Open the *Mineral* folder and go to the *Gold Mine* placemark to see a representative scene of a large-scale gold mining operation. Mining operations can have major impacts on the local environment and nearby landscapes through the mining process and associated infrastructure. Analyze the site and surrounding area near the *Gold Mine* placemark.

Exploration 18.3 MULTIPLE CHOICE

1. Based on your analysis of the *Gold Mine* site and the surrounding area, identify the statement that is most strongly supported.

    A. This is a remotely operated, unmanned mining site.
    B. Mining is most likely happening primarily underground.
    C. Surface water at the site has a variety of colors, suggesting a variety of geochemical processes and/or components.
    D. It appears that rail is the primary transportation mechanism from the site.
    E. It appears that there is no connection from the mine to the sea via pipelines

2. Another precious metal is platinum. Fly to the *Platinum Mine* placemark to see a representative mine in South Africa. The vast majority of platinum is mined in South Africa and the majority of platinum is used in one industry. Using the Internet and other resources, what placemarked city, based on their predominant industry, is most dependent upon mines like this one?

    A. A
    B. B
    C. C
    D. D
    E. E

Go to the *Bingham Canyon Mine* placemark to see a famous open-pit copper mine. Turn on the *Photos* layer and view some of the pictures of the site. This site is actually on the National Register of Historic Places (NRHP). You can look up NRHP information from the linked search page.

3. Identify the statement regarding the Bingham Canyon Mine that is supported by NRHP data.

    A. The owner of the property is American Copper Company.
    B. Copper production at the mine is expanding and will continue until at least 2017.
    C. The mine is more than five miles deep.
    D. Material from the bottom of the mine pit is removed by helicopter.
    E. The mine yields more than 98 percent of the copper produced in the United States.

Open and turn on the *Metallic Mineral* folder. Five countries are placemarked. These represent the top five producing countries for a particular metallic mineral. Because metallic minerals are so unevenly distributed around the world, it is important to have standardized information regarding their distribution. This is provided by the US Geological Survey (USGS). Open the link associated with the *Metallic Mineral* folder to see a list of minerals and materials about which the USGS maintains statistics.

4. Utilize 2011 USGS data to determine which metallic mineral is represented by the top five producers that have been placemarked in the *Metallic Mineral* folder.

   A. chromium
   B. copper
   C. silver
   D. tungsten
   E. zinc

Exploration 18.3 Short Essay

1. Go to the *Minerals Here* placemark. Complete some outside research to determine the answers to the following questions: What are some examples of minerals found here? Are they currently being mined? Why or why not?

   _____
   _____
   _____
   _____
   _____
   _____
   _____

2. Using Google Earth™, find a mineral resource extraction site near you. Describe the site. Use your detective skills to find out what is being extracted there. What products rely on this material? If you are having trouble locating a site with the imagery alone, try typing the word "mine" into the Find Business tab of the Search pane.

   _____
   _____
   _____
   _____
   _____
   _____
   _____
   _____

Exploration 18.4: Biotic

The biotic, or living, resources of the planet provide humans with important services, as well. These ecosystem services are not as easily quantified as an ounce of gold, for example, but they are every bit as valuable. The forests of our planet not only provide us with wood for building and paper products, but are also vital habitat for our planet's fauna and provide essential services to our atmosphere by absorbing carbon dioxide. The impacts of logging are readily seen in Google Earth™ using historical imagery. Some environmental groups have paired imagery with information they have added to the Google Earth™ environment. One example is presented by the *Neighbors Against Irresponsible Logging* group. To see the Google Earth™ layer the organization created, turn on the *Santa Cruz Mountains Redwood Logging Plan* folder. Explore the nested folders to see what kinds of information they included in their "activist" Google Earth™ project. When you have completed the question, turn off and collapse the *Santa Cruz Mountains Redwood Logging Plan* folder.

Exploration 18.4 MULTIPLE CHOICE

1. After exploring the contents of the *Santa Cruz Mountains Redwood Logging Plan* folder, select the statement that is most strongly supported.

   A. Three communities are potentially affected by the logging plan.
   B. Los Gatos Creek can be described as a largely concrete urban stream.
   C. Logging would potentially occur within a few hundred yards of several schools.
   D. The area to be logged stretches more than 15 kilometers from end to end.
   E. The area to be logged surrounds the city of Santa Cruz.

Turn on the *CONABIO COP16 Tour*. Expand the folder and then click the "click to play tour" link. Be sure you have the volume turned on. When you have answered the following question, turn off the *CONABIO COP16 Tour* folder.

2. After watching the mangrove forest tour, identify the statement that is most strongly supported.

   A. Mangrove forests are centered on mountainous areas.
   B. Mangroves lack significant species diversity.
   C. Human activities pose the biggest threat to mangrove forests.
   D. Shrimp farming has benefitted mangrove forests.
   E. Mangrove forests are only found on the west coast of Mexico.

Turn on the *ARKive: Endangered Species* folder. Expand the folder to see a list of placemarked endangered species. Use the placemarked species to learn more about the status of some of the world's endangered species. When you have completed the following question, turn off the *ARKive: Endangered Species* folder.

3. Utilizing the *ARKive: Endangered Species* folder, identify the statement that is most strongly supported.

   A. The takahe is threatened because it cannot fly as fast as its predators.
   B. Wolverines have suffered from human disturbances to the remaining areas of wilderness where they live.
   C. The snow leopard is endangered because it is only found high in the Colorado Rocky Mountains.
   D. There are no critically endangered species in the United States.
   E. Bristlecone pines are vulnerable because they can only live to be 200 to 300 years old.

4. Turn on the *Coral Reef Watch* folder. The National Oceanic and Atmospheric Administration (NOAA) logos indicate reporting stations utilized for the monitoring of coral health. This provides you with a rough approximation of the location of coral reefs. The distribution of coral reefs is largely impacted by water temperature. What region/area does not have any reef monitoring stations?

   A. Florida
   B. East Africa
   C. Persian Gulf
   D. US West Coast
   E. Indian Ocean

Exploration 18.4 SHORT ESSAY

1. Double-click the *Bleaching Alert Area* folder in the *Coral Reef Watch* folder. This shows areas where reefs are susceptible to bleaching as a result of elevated sea temperatures. You can also use the *Bleaching Outlook* folder. What is coral bleaching? What areas are most at risk? What do you think the layer would like six months from now and why?

_____
_____
_____
_____
_____
_____
_____

2. In this exploration, you have seen four applications of Google Earth™ technology leveraged to enhance environmental awareness and understanding. Which one did you think was most effective in achieving its goal and why?

_____
_____
_____
_____
_____
_____
_____

Name: _____

Date: _____

Chapter 19:
Nature and Society

Human-environment interaction is a complex, constantly evolving, two-way dialogue. Humans are presented with varying environmental situations and challenges that can result in a wide range of actions. Many of the choices we make have profound implications on the planet's air, land, water, and living organisms. In this exploration, you will analyze some examples of human-environment interaction, including the concept of conservation, the problems of pollution, and the wide-ranging concerns associated with climate change.

Download EncounterHG_ch19_Nature_and_Society.kmz from **www.mygeoscienceplace.com** *and open in Google Earth™.*

Exploration 19.1: Human Environment Interaction

While the news of today often contains stories of the disharmony between human populations and the "natural world," this is not a new phenomenon. In fact, humans have been impacting the environment for thousands of years through the manipulation and exploitation of resources. Open the *Human-Environment Interaction* folder followed by the *Paleolithic Environmental Impact* folder. Fly to the three sites and survey the landscapes. You may also find it useful to turn on the *Photos* layer.

Exploration 19.1 MULTIPLE CHOICE

1. What environmental impact is represented by the sites featured in the *Paleolithic Environmental Impact* folder?

    A. erosion from intensive farming
    B. manipulation of animal populations through hunting
    C. pollution of groundwater from excessive fertilization
    D. burning to control growth of trees
    E. importing non-native crops

When Europeans arrived in the Americas, a process referred to as "ecological imperialism" occurred. Research this term and then open the *Ecological Imperialism* panoramic image.

2. What element of the *Ecological Imperialism* panoramic image best illustrates the concept of ecological imperialism?

    A. the combine
    B. the big bluestem grass
    C. the row patterns in the field
    D. the wheat in the field
    E. the power lines

Not all of the exotic plants and animals that have been introduced have been beneficial. At times, invasive species will find a niche in their new environment and flourish. Turn on the *Invasive Species* folder to see an overlay of states that have been impacted by a quickly growing exotic. The vine, native to southern Japan and southeast China, has been called the "vine that ate the South," as it has expanded rapidly since its first documented introduction to the United States in 1876. A great place to find information about invasive species is the United States Department of Agriculture's National Invasive Species Information Center. When you have completed this question, turn off the *Invasive Species* folder.

3. The *Invasive Species* folder illustrates the distribution of what invasive plant?

    A. Kudzu
    B. Russian Olive
    C. Purple Star Thistle
    D. Diffuse Knapweed
    E. Bloodrot

To see one of the most striking examples of human impact on the environment, fly to the *Aral Sea* placemark. The Aral Sea used to be one of the largest lakes in the world. Today it is a fraction of its original size because of river diversions for irrigation projects in the former Soviet Union. Examine the site in Google Earth™ using the historic imagery capabilities.

4. Based on your analysis of the Aral Sea imagery, select the statement that is most strongly supported.

    A. The lake appears to begin its desiccation in 1999.
    B. The remnants of the lake are located in Uzbekistan and Turkmenistan.
    C. The now dry lakebed has become an area of intense cultivation.
    D. The residual salt from the lake's desiccation is clearly visible in the imagery.
    E. There are no communities within 100 kilometers of the Aral Sea.

Exploration 19.1 SHORT ESSAY

1. Research the Aral Sea situation further utilizing your textbook and outside resources. Also turn on the *Photos* layer and explore images of the area. What are some of the hazards and economic problems faced by people of the region as a result of this human-made environmental catastrophe?

_____

_____

_____

_____

_____

_____

_____

2. Go to the *Mississippi Delta* placemark to see an area that bears the unmistakable imprint of human activity. What are these linear features and why are they here? Why should there be concern that the presence of these features leads to erosion that destroys the delta and wetlands? Draw upon outside resources to help you answer these questions.

_____

_____

_____

_____

_____

_____

_____

Exploration 19.2: Conservation

As humans have witnessed the despoliation of the natural environment, there have been large numbers of individuals, groups, and movements that have arisen in the name of conservation or preservation. In the United States, the conservation movement was bolstered in the nineteenth century by the reports of explorers from the American West. One important expedition group was the Hayden Geological Survey of 1891. This group explored the Yellowstone region and documented the area with photographs, paintings, and detailed scientific journals. One of these paintings, titled "Grand Canyon of the Yellowstone," provoked residents of the eastern United States to champion preserving this special place. Go to the *Grand Canyon of the Yellowstone* placemark to see the general vantage point from which this painting was based. Then research the painting on the Web to see it.

Exploration 19.2 MULTIPLE CHOICE

1. After examining the painting and the Google Earth™ perspective, identify the statement that is most strongly supported.

    A. The Google Earth™ perspective shows that a large hotel complex has been built at the site.
    B. There is little to no difference in the Google Earth™ perspective and the painting.
    C. The Google Earth™ perspective illustrates the geysers of Yellowstone more than the painting.
    D. More of the river is visible in the painting.
    E. Moran's painting includes people.

The documentation of the Hayden expedition was instrumental in the establishment of Yellowstone National Park. The US National Park System has served as a model for the world with many of our most important natural and cultural resources granted a measure of protection for the enjoyment of this and future generations. Expand the *More* folder in the *Primary Database*, followed by the *Parks/Recreation Areas* folder, the *US National Parks* folder, and the *Park Descriptions* layer. Then, zoom out to the point that you can see the 48 contiguous states. The hiker icons illustrate the distribution of the national parks with a focus on natural attributes. You can click on the icons to learn more about the attractions at each site. Turn off the *Park Descriptions* layer when you complete the following question.

2. Based on your assessment of the national parks with a natural emphasis, select the statement that is most strongly supported.

   A. The eastern United States has more national parks than the West.
   B. There are no parks that highlight caves.
   C. The Great Plains area of the United States is underrepresented in the National Park system.
   D. Coral reefs are not represented in the system.
   E. Colorado has the most national parks with a natural theme.

The ocean's resources have increasingly become a target of conservation in recent years as fisheries around the globe have decreased in productivity or altogether collapsed. Turn on the *Global Fisheries 1950–2004* folder. Read the introduction to the project associated with the placemark near Vancouver. then zoom out and start exploring the layer.

3. Identify the statement that is most strongly supported by the data contained within the *Global Fisheries 1950–2004* folder.

   A. In the West Central Atlantic zone, catches steadily increased from 1980 to 2004.
   B. Many of the fisheries west and south of Hawai'i displayed sharp declines in the 1990s through 2004.
   C. In the Southwest Atlantic zone, the top fishing countries are from Europe or North America.
   D. All fisheries around the globe have shown decline in catches from 1950-2004.
   E. In the Pacific Northeast zone, the top species caught from 1950 to 2004 was salmon.

Water supply forecasts are essential to managing the crucial resource of water. The US Department of Agriculture's Natural Resources Conservation Service (NRCS) carefully monitors water supply in the semi-arid western United States. Turn on the *Water Supply Forecasts* folder. You will see a series of color-coded cones indicating current water forecasts for the major hydrologic basins of the western United States. Click on any cone to see a snapshot of the forecast for that river. For a more comprehensive assessment, click the *State Basin Outlook Report* link in the upper-left corner.

4. Which of the following is not a primary component of the NRCS's state basin outlook reports?

   A. agricultural production
   B. snowpack
   C. precipitation
   D. streamflow
   E. reservoirs

Exploration 19.2 SHORT ESSAY

1. Find a conservation-related project in or near your community that is visible in Google Earth™. Perhaps there is an area of wetlands restoration, brownfield conversion, urban forestry, or erosion control? Describe the scene and the purpose of the project.

   _____
   _____
   _____
   _____
   _____
   _____
   _____

2. Utilizing the water supply forecasts, provide a brief regional analysis of the water situation in the western United States. What areas/states are forecast to be wettest and driest as of today's current date? Turn off the *Water Supply Forecasts* folder when you have completed the question.

   _____
   _____
   _____
   _____
   _____
   _____
   _____

Exploration 19.3: Pollution

Unfortunately, one of humankind's most striking contributions to the human-environment dynamic is its ability to pollute. For thousands of years, humans have fouled the air, water, and land of the planet. One form of pollution that has started to receive increased attention is the problem of marine debris. Open the *Marine Debris* folder and work your way through the placemarks from top to bottom to develop your knowledge of the various environmental issues that are currently occurring in the oceans.

Exploration 19.3 MULTIPLE CHOICE

1. Identify one of the emerging concerns associated with marine debris.

   A. Microplastic ingestion
   B. Ocean cooling resulting from reflection of sunlight by debris
   C. Ships sunk by pirates
   D. Dumping of fishing gear by Koreans
   E. Gear dumped by scuba divers

When it comes to major pollution concentrations in urban areas, there is often a strong socioeconomic correlation. The concept of "NIMBY," or "not in my backyard," is manifested by the greater political and economic power for more affluent neighborhoods to steer projects such as the city dump away from their neighborhoods. Oftentimes, such facilities end up being located in a part of the city with marginal political power and limited economic resources to pursue litigation. For example, go to the *Little Rock Dump* placemark. To get a feel for the socioeconomic conditions in this part of the city, click the link to ESRI's Zip Code Look-Up tool.

Type the zip code 72204 that contains the Little Rock dump into the "Try another ZIP Code" search box.

2. Identify the statement that is most strongly supported by the zip code data from ESRI for zip code 72204.

    A. Unemployment in this zip code is historically lower than the national average.
    B. Household income is less than $50,000 for more than 60 percent of households.
    C. The greatest percentage of jobs in this zip code can be described as blue collar.
    D. There are more men in this zip code than women.
    E. This zip code is predominantly Hispanic.

La Oroya, Peru, is one of the most polluted places on the planet. According to *Time* magazine, more than 99 percent of children in the town have levels of pollution in their blood that exceed suitable limits. Survey the scene and use the Internet to research the cause of this pollution.

3. Identify the most suitable term associated with the pollution in La Oroya.

    A. chemical weapons
    B. nuclear
    C. fertilizer
    D. petrochemical
    E. smelting

One realm of pollution that rarely gets much attention is light pollution. Light pollution obscures the stars in the night sky, interferes with astronomical observatories, and disrupts living systems. Disruptions are centered on plant and animal physiology that can lead to the alteration of entire ecosystems. In recent years, the dark-sky movement has worked to reduce light pollution. Nonetheless, many parts of the globe have nighttime brightness levels that are more than 100 times the natural level. A good way to see this is to turn on the *Earth City Lights* layer in Google Earth™. In the *Primary Database*, open the *Gallery* folder, then open the *NASA* folder, and turn on the *Earth City Lights* layer. When you have completed this question, turn off the *Earth City Lights* layer.

4. What country appears to have the lowest levels of light pollution on average?

    A. Slovenia
    B. Qatar
    C. Sri Lanka
    D. French Guiana
    E. South Korea

Exploration 19.3 SHORT ESSAY

1. Use the ESRI Zip Code Look-Up tool to assess the demographics of an area that you associate with higher pollution in your community. Report your findings.

_____
_____
_____
_____
_____
_____
_____

2. Air pollution is monitored by the Environmental Protection Agency (EPA) in the United States. The EPA Air Quality Index (AQI) "identifies air quality as it relates to health effects you may experience within a few hours or days after breathing polluted air." Turn on the *Air Quality from AIRNow* folder. Find the monitoring site nearest your current location and report the conditions. Then look for the highest readings you can find and report the conditions. Turn off the *Air Quality from AIRNow* folder when you complete the question.

_____
_____
_____
_____
_____
_____
_____

Exploration 19.4: Climate Change

Perhaps the defining issue of human-environment interaction in the twenty-first century is climate change. While most people in the developed world may not directly feel the impacts of changing climatological regimes at the moment, the same cannot be said for many people in the developing world. As governments grapple with the issue, achieving consensus on how to address the problem continues to be difficult. For example, should developing countries face sharp curbs on their carbon emissions if the developed countries of the world had no such limitations as they developed? Open the *Climate Change* folder and then open the *Carbon Dioxide Emissions* folder to see the world's top 20 emitters. Darker shades of blue represent higher values. Turn on the *Carbon Dioxide Emissions Per Capita* folder to see the same countries' values on a per capita basis. Again, the darker shades of blue represent higher values. Utilize these datasets to respond to the following question. When you have completed the question, turn off the folders.

Exploration 19.4 MULTIPLE CHOICE

1. Utilizing the *Carbon Dioxide Emissions* folder and the *Carbon Dioxide Emissions Per Capita* folder, identify the statement that is most strongly supported.

    A. Indonesia is in the highest category for carbon dioxide emitters on both a total amount and per capita basis.
    B. Italy appears to have the largest gap between total emissions and per capita emissions.
    C. While India and China are both large emitters of carbon dioxide, they do not emit as much as the United States on a per capita basis.
    D. No African countries are in the top 20 emitters of carbon dioxide.
    E. While Australia is a significant emitter of carbon dioxide, it has a relatively low per capita value.

2. Turn on and expand the *Climate Change KML Tour* folder. Start the tour as directed in the folder. Be sure you have the volume on your computer turned up. Identify the statement that is *not* supported by the National Snow and Ice Data Center climate change tour. When you complete the tour, turn off the *Climate Change KML Tour* folder

    A. Most of the glaciers in the world are retreating.
    B. Antarctica is usually more sensitive to climate change than the Arctic.
    C. The Arctic ice sheet reaches its maximum extent in March.
    D. Thawing permafrost releases large amounts of carbon dioxide and methane.
    E. Scientists consider the globe's polar regions to be like "a canary in the coalmine" in terms of the effects of climate change.

Turn on and expand the *IPCC Climate Projections, 2004–2100* folder. Click the "Click here to play tours" link, and then click the "high emission scenario *Temperature*" link. Study the distribution of increased temperatures over time. Keep in mind this is the Celsius temperature scale.

3. What city is projected to experience the greatest change in temperature under the high emissions scenario?

    A. Buenos Aires
    B. Vancouver
    C. Bangkok
    D. Baghdad
    E. Belgrade

Now watch the "high emission scenario *Precipitation*" animation. Evaluate the data and keep in mind that the scale is in millimeters per day.

4. Select the statement that is most strongly supported by the average precipitation change under the high emission scenario.

   A. Virtually the entire globe will be drier under this scenario.
   B. Indonesia looks to become much wetter.
   C. Middle America will become wetter in the coastal areas.
   D. Virtually the entire globe will be wetter under this scenario.
   E. The United States will become wetter in the South and drier in the North.

Exploration 19.4 SHORT ESSAY

1. After watching the National Snow and Ice Data Center climate change tour, what illustrative element did you find most effective in explaining the significant changes taking place in the polar regions?

_____
_____
_____
_____
_____
_____
_____

2. Utilizing the high emission projections from the IPCC, describe the changes that could be experienced at your location.

_____
_____
_____
_____
_____
_____